## An Inspiring Account: A Rich, Exciting Story

*The Spirit of Texas* is a historical family insight of one state's reflection of American exceptionalism. Raw, rugged men and women of noble character building a nation from grit, courage, and compassion. A free, industrious people, unyielding to defeat and seeing their cut of cloth as identical to our Framers, who were independent, self-reliant, and God fearing. The Texas spirit is still alive today, and that Lone Star still carries the weight of fifty not bowing to perverse ideology but acknowledging the laws of nature and nature's Creator. That spirit still stands today as a beacon of light in our schools, churches, centers of commerce, and government for a whole nation to draw upon.

John Sauers, Tea Party Founder
Loganville, Georgia

**Bravo!** *The Spirit of Texas* tells the *truth* about the real Texas and the people who first settled it. Being raised in this ranching territory as a descendant of ranching families, I can testify that this is a genuine story of a west Texas pioneer family that managed to succeed in spite of many hardships. Through these difficulties and their love for the Lord they became stronger. The spirit of Texas is still alive and well today in this great state because of our ancestors' true grit and independent spirit. **God bless Texas!**

Lee Weddell Puckitt
Immediate Past President
Texas Sheep and Goat Raisers Association

*The Spirit of Texas* is the story of one Texas family's heritage and the building of their ranch as the great state of Texas was also being built. It is a wonderful personal story that will engage and delight Texans young and old, and it's a perfect supplement to Texas history for students and adults alike. It is not one of those histories that has been *sanitized* to remove any reference to our ancestors' Christian faith, but it faithfully presents life as Texans lived it.

Tim Lambert
Texas Home School Coalition

Among the sons of Scotland are famous explorers and botanists who visited faraway lands, including one who was the first European to ascend Hawaii's mighty Mauna Loa and another was first to the North Pole. They discovered new plants and animals and performed grand experiments. William Menzies, the subject of this biography, called *The Spirit of Texas*, was born in Aberdeen, Scotland, and as a baby immigrated to Canada with his family in 1856. He settled in west Texas, where he met and married Letha Ann Chastain in 1888.

*The Spirit of Texas*, written by his great-grandson Winston Menzies, records how his great-grandparents survived against all manner of trials, prospered, and raised eight children on a remote Texas ranch by the San Saba River. They succeeded by their faith in God, hard work, and dedication to family and to one another. The Menzies story is an intriguing, adventuresome reminder of the pioneering spirit that made rural Texas so unique. It is also an admonition of how we have abandoned so much of what made this nation possible. As the author observes in his epilogue, "If we lose the history of where we came from, we certainly won't know where we are going. God help us to at least leave it the way we found it and, hopefully, by His grace, make it even a little better."

*Forrest M. Mims III, "The Country Scientist"*
*Writer for scientific journals and the* San Antonio Express-News
*Named one of the "50 Best Scientists" by Discover Magazine*

*The Spirit of Texas* is not only filled with good stories of the Menzies family from the time they arrived in Menard County in 1887 when the Hill Country was still open range, it also gives a detailed history of the Presidio de San Saba and Jim Bowie era. I have known many of the family members over the past 50 years and it was hard to put this book down.

*Jerry Lackey*
*Agricultural Journalist*
*San Angelo, Texas*

## Other Books and CD Series by the Author

### Being God's Man
*Booklet and Four Audio Presentations*

### Called to Lead
*Three Audio Presentations*

### How to Build a Debt-Free Home
*Booklet and Three Audio Presentations*

### How to Buy a Used Car

### Prudence
*Ten Audio Presentations*

### Secrets of Prayer That Avails Much
*Booklet and Four Audio Presentations*

### The Fulfilled Life in Christ
*Book and Four Audio Presentations*

### The Success Secret of the Ages

### Total Victory
*Four Audio Presentations*

### Walking in Divine Favor
*Booklet and Four Audio Presentations*

WinstonMenzies.com

Biography of William Menzies

# THE Spirit OF TEXAS

*The Astonishing Story of a Pioneer Rancher's Family and Their Mighty State*

by
## WINSTON MENZIES

CREATIVE PUBLISHING COMPANY

THE SPIRIT OF TEXAS
Published by Creative Publishing Co.
P.O. Box 90
Conyers, Georgia 30012
WinstonMenzies.com
CreativePublishing.us
SpiritofTexas.us

Published with assistance from Creative Enterprises Studio.

Copyright © 2011 Winston Menzies

All rights reserved. No part of this book may be reproduced or transmitted in any form or by any means, electronic or mechanical, including photocopying and recording, or by any information storage and retrieval system, without permission in writing from the publisher, except for brief quotations in critical reviews and articles.

All Scripture quotations are taken from the New King James Version. Copyright © 1982 by Thomas Nelson, Inc. Used by permission. All rights reserved.

Cover design by Michael Gore, It's Just a Pixel, Fort Worth, Texas, www.itsjustapixel.com

Library of Congress Cataloging-in-Publication Data

Menzies, Winston.
  The spirit of Texas : biography of William Menzies, west Texas rancher / Winston Menzies.
    p. cm.
  Includes bibliographical references.
  ISBN 978-0-9837472-0-8 (tradepaper)
  1. Menzies, William, 1855-1957.  2. Ranchers--Texas--Menard County--Biography.  3. Immigrants--United States--Biography.  4. Menard County (Tex.)--Biography.  I. Title.
  CT275.M46934M46 2011
  976.4'87706092--dc23
    [B]
                                                            2011031348

Printed in the United States of America

11 12 13 14 15 TS  5 4 3 2 1

*To my loving and faithful parents,*
*Col. Perry Phillip and Mary Louise Menzies,*
*true Texans in every way.*
*No one could ask for better parents.*
*I am indebted to them in so many ways; they are innumerable.*

# CONTENTS

| | |
|---|---:|
| *Acknowledgments* | xi |
| **Menzies Family Tree** | xiii |
| 1  The Early Days | 3 |
| 2  Going West | 9 |
| 3  Settling in Menard County | 21 |
| 4  Letha Ann | 37 |
| 5  Character of These Pioneers | 57 |
| 6  Ranching in West Texas | 63 |
| 7  Pioneer Days | 79 |
| 8  Progressive Rancher | 151 |
| 9  A Blessed Union | 189 |
| 10  The Latter Days | 211 |
| *Epilogue* | 245 |
| *Texas State Resolution No. 68* | 261 |
| *Bibliography* | 263 |
| *Index* | 267 |
| *About the Author* | 274 |

# ACKNOWLEDGMENTS

THIS ACCOUNT OF THE lives of William and Letha Ann Menzies was made possible by the generous and loving input of all my cousins, especially Willie Lee Everett. She has been the spark plug of our clan for years and a joy to everyone who has been blessed to know her. Initially, it was her giving me written materials and encouragement that led me to write this book. She was also a delight to talk with almost daily over the four years I was investigating, accumulating information, and writing. She lovingly tutored me, at a young eighty-seven years of age, with her knowledge of Menard like one of the thousands of students she taught in the public schools over her lifetime. They all loved her too.

To my wonderful wife, Donna, who was a daily help and encouragement. To my two sons, Ron and John, for their help. To my many relatives who have generously shared their memories, accounts, and pictures as well as helped do a great deal of research in various areas. It was also made possible by the written accounts of the late Ann and Kitty Sue Menzies. Their contributions, in turn, were in large part made possible by the input of the late R. J. Godfrey, Max and Alex Menzies, as well as Willie Lee's late husband, Allan Everett. Many thanks also to Margo Seaman for her excellent photos of the Crawford and Menzies ranches and to Gary Cutrer, editor of *Ranch & Rural Living* magazine, for the pictures they shared.

Many thanks also to John and Katherine Kniffen. To John for the depth of his lifelong study of the native Indian tribes of Texas and beyond and for his gracious spirit, sharing it all with me. To Katherine for her willingness to share freely all the artifacts of the Menard Historical Society, of which she is president.

When I think about the accomplishments, many kindnesses, and the generosity of William and Letha Ann Menzies, not forgetting their humility, it is clear that they were giants in their day, and I find myself an unworthy but grateful descendant of this noble ancestry. It is well to remember the many sacrifices and hardships of those who have gone before us.

Without question, their sacrifices and contributions have made our lives and circumstances possible. More important, it was from them that we have derived our character, our Christian heritage, our work ethic, and our good name. For all these things and much more, we will be forever indebted.

# Menzies Family Tree
## 1820 – 1950

**William Menzies Sr. & Agnes (Craigmile) Menzies**

- **William & Letha Ann (Chastain) Menzies**
  - George & Ella (Phefer) Menzies
    - Perry & Mary Louise Menzies
      - G. C. & Estelle Menzies — Craig, Mike, Tyra, Nancy Menzies
      — Steve, Winston Menzies
  - Raymond & Agnes (Menzies) Walston
    - Raymond Roy & Jaine Walston — Ann, Lee, Walker, Roy Jr. Walston
    - Allen & Willie Lee (Walston) Everett — Carolyn, Pat Everett
  - Bill & Anne (Crawford) Menzies
    - Ted & Billie Anne (Menzies) Cannon — Cynthia, Pam Cannon
    - Don & Marie (Menzies) Scruggs — Margo, David, Kevin Scruggs
  - Tom & Letha (Menzies) Jacoby
    - Ray & Jean Jacoby — Janna, Randa Jacoby
    - Roy & Ann Jacoby — Jamie Roy, Nancy, Jaine, Letha Jacoby
    - Phillip & Lynn Jacoby — Mark, Scott Jacoby
  - Irby & Pearl (Menzies) McWilliams
    - A. J. & Louise McWilliams — John McWilliams
    - Arel & Mary Pearl Faver
    - M.D. & Sidney McWilliams — M.D. Jr. Calvin, Irby McWilliams

- **Alex & Mary Menzies (No Children)**

- **George Menzies**
  - Alex & Marguerite (Watson) Menzies
    - A.L. (Sonny) Menzies Jr. — Stephen, Linda Menzies
    - Carl & Shirley Menzies — Don, Mark Menzies
    - Jim Joan Menzies
  - Max & Kitty Sue (Harrison) Menzies
    - Duery & Lawanda Menzies — Misti, Dusty Menzies
    - William (Duck) & Jenna Menzies — Wil, Dawn, Grant Menzies
    - John & Brenda Menzies — Leslie, Ashley Menzies
  - Walter & Hazel (Whitley) Menzies
    - Walter Scott (Scotty) & Lucille Menzies — Wes, Ward, Wade Menzies

# Chapter 1

# The Early Days

**O**VER TIME, THE SPIRIT OF TEXAS rested on countless rugged, pioneer men and women too numerous to name whose brave acts of courage and selfless duty won the West. They also gave birth to future generations of stalwart citizens just like themselves. These pioneers met the dangers and difficulties of frontier life head-on and subdued them. Their individual acts of bravery and devotion have often gone unnoticed or have, regrettably, long since been lost in the sands of time. Remembering our heritage

Castle Menzies

and learning the lessons our ancestors so ably taught us keep future generations from having to continually start over from scratch. Simply put, pioneering Texas was like riding a wild bronc that had to be broken. You will either have enough of the spirit to break it, or it will break you. It has been said (and rightly so) that Texas was hard on men and dogs but hell on women and horses. For the most part it is not a gentle land. Although beautiful in its own right, it was and remains an arid, unforgiving land. The farther west you go, the terrain becomes more and more a land of scrub oaks, mesquite trees, cactus, rattlesnakes, jackrabbits, armadillos, and scorpions. In addition to all this, the early settlers had to deal with Indians, outlaws, and the fact that Spain and subsequently Mexico wouldn't allow them to own even a handful of this Texas dirt until 1821. Even then, landownership came with onerous, unbearable conditions. Without an understanding of the struggle this land inspired and required, Texas will mean nothing to you.

This is a true story, gleaned from documents recently found, of a Scottish immigrant who crossed the great ocean with his family to forge a new life in North America. He came first to Canada, then to New York, and finally to Texas at the age of twenty-one with nothing but a wooden tool box. Here he built a family and a legacy of self-reliance and dogged determination, carving out a beautiful but hard-won life as a pioneer rancher. He also became the patriarch of the Texas Menzies clan, and through his character, forged a singular heritage for generations to come.

William Menzies arrived at the port city of Aberdeen, Scotland, on July 29, 1855, the firstborn son of William Sr. and Agnes Craigmile Menzies, natives of Scotland. Later the family moved to Inverness, from which they set sail for America a year or so later. In those days, the Menzies were about as plentiful in Scotland as the Smiths and Johnsons are here in the United States today.

The family lived very near Castle Menzies, which had been the seat of the chiefs of the Menzies clan for over four hundred years. The castle is at Weem in Perthshire, which is about a mile from Aberfeldy and in the middle of the very turbulent history of the Highlands. This was where Bonnie Prince Charlie took a brief respite on his way to Culloden in 1746. Having been schooled in military science from his youth for this purpose, he was sent to Scotland by his father, James VIII, who was in exile in Rome, to start an uprising to resecure the English Crown for himself. Charlie initially met with considerable success in securing Scotland, but he was soon soundly defeated by the duke of Cumberland, dubbed "The Butcher," at Culloden. He ultimately returned to Rome and lived in exile for the rest of his life as well. It was widely believed that his Catholic background kept many of the English Protestants from supporting his cause for fear he might restore Catholicism as the national religion. Charlie's defeat brought on years of fierce persecution for the Scots and oppressive laws that forbade the Highlanders from carrying weapons, bagpipes, or wearing the tartan or the kilt.

Menzies coat of arms

The Menzies clan's motto of old, which can be found on their coat of arms in Scotland today, is from the old English, *Vil God I Zal,* which in today's vernacular means "Will God I shall." In other words, if it is God's will, it will be so and happen in our lives regardless of the obstacles, impossibilities, or difficulties that might arise. God is working at all times on this planet to bring about things according to the counsel of His will, and our lives are not our own. He "has determined their preappointed times and the boundaries of [our] dwellings" (Acts 17:26). We have seen how God has had His fingerprints all over this family as far back as we have any recorded information. It was God's will that this small family of Menzies would come to the New World, as it was considered then, by His hand and will. It might have been thought at the time that it was more for other reasons, such as the scarcity of land or persecution in Scotland or greater opportunity in a new, burgeoning country, but it was, nonetheless, by God's hand.

William's dad was a builder, and there happened to be a great demand for his type of work in North America. Alexander, one of his uncles, was a master carpenter while the other uncle, John, was what was called an "art master." One of William's uncles in Scotland was the captain of a sailing ship who later went down with it in a storm at sea. His parents, William Sr. and Agnes, had seven children, but only three boys reached maturity: William, Alex, and George. Of the others about which we know anything, Frank lived less than a year and died in 1874. Charles lived less than a month, dying in 1874, and James was twenty-one when he died in 1881. A daughter named Agnes lived only three months, passing in 1869. One son, Duryee, died of lockjaw in Tarrytown, New York, in 1881, when he was fourteen years old.

## Setting Sail for a New Country

Eighteen months after William was born, his parents gathered up the family and set sail for Canada. The trip across the ocean lasted six weeks, and they encountered at least one bad storm during the voyage. According to his parents, William was a good sailor, as he and his father stayed on deck mostly, and he never got seasick. At the time he was just learning to talk, but William quickly took up with the sailors and could say "heave ho" with the best of them. The family first sailed down the St. Lawrence, landing in Hamilton, Canada in 1856, and then, in 1860, when William was five years old, they entered the United States. First settling at Glenham in Duchess County, they moved to Fish Kill Village and later to Tarrytown, where William continued in his building trade. They finally established a home at Irvington on the Hudson River in New York State, and this was where William Jr. received his education. His father died in 1899, and his mother passed away only a year later, in 1900.

William learned the stone and brick mason trade from his father at an early age. As a kid, he loved to help his dad lay bricks. He was a quick study and had a quick wit. One time his dad was working with an Irishman named Pat (or Paddy), an immigrant from Cork. The man came to work one morning with a big hole in his britches. Young William quickly made up a little rhyme about it: "Paddy from Ireland, Paddy from Cork,

*The Early Days*                                                                                      7

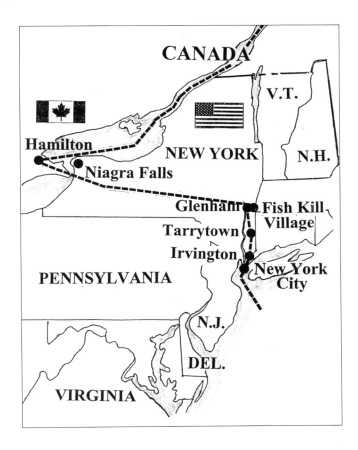

Paddy had a hole in his britches as big as the state of New York." Well, Pat didn't cotton to it too well and later, whether intentionally or not, a brick was dropped from a scaffold and landed on William's head. It made a three-cornered impression on his forehead that he carried for the rest of his life.

William also learned the carpentry trade from his father and started working in the shop when he was about nineteen years old. His first job was carrying bundles of shingles that were to be cut by running them through the saws. The machine was operated by a steam-powered engine and muscle. Unknowingly, certainly at the time, young William was developing an aptitude and a love for machinery. The shingles were often frozen together, so he had a hard job just getting them separated. He stayed in the machinery section of the shop until he and a friend, Andy Elder, started building greenhouses throughout the Northeast. These greenhouses were helpful for starting plants and flowers during the hard, cold winters prevalent in that section of the country. Young William and Andy continued working in the construction trade, building greenhouses for several years, even to include a large one at the Roosevelt compound at Hyde Park, which later became the home of Franklin D. Roosevelt. William and Andy were enjoying the building business and were making pretty good money at the time: $2.50 a day.

While working in Massachusetts, the two young men took a trip to Plymouth Rock and stood on the spot where the Pilgrims had landed. William said the big black rock was about fourteen feet wide and twenty feet long.

## Chapter 2

# Going West

AT THE AGE OF twenty-one, William became a US citizen. In that same year he was struck with a strong case of wanderlust. There was a spirit that spoke to so many young men in those days, saying, "Go west, young man. Go west." And indeed William did. Without telling anyone where he was going, he boarded a ship bound for Galveston, Texas. He did, however, take the time to write back home shortly after his arrival. He bought a steerage ticket on the Mallory Line on the *Lukenbach*. He couldn't afford the best accommodations, so he slept on the deck. The food wasn't the greatest either, so he gave the cook two dollars extra to let him eat with the petty officers. Much later in life, in his early nineties, he was asked why he left so abruptly and came to Texas. He laughed and said, "I left New York because, just like a Scotsman, I wanted to prowl around. They say, you

William's tool chest and key

know, that when they discovered the North Pole, they found a Scotsman there astraddle it."

William arrived in Galveston in June 1876 with nothing but his tool chest. His hope was to find work as a carpenter, but that was next to impossible because, at the time, all the carpenters in Galveston spoke German. William spoke not a word of German, and he had no intention of doing so. He arrived, however, at the time of one of the great Galveston hurricanes and floods. The floodwaters during one hurricane were increasing to the extent that the water rose in their shop, and another man set his tool chest on top of William's. Fortunately, William found it in time to right the situation and save his tools. Since the box and most carpentry tools were wooden in those days, getting them soaked would have completely destroyed them. Galveston was also no stranger to fires, and an enormous fire had recently swept through the area. Much life and property were lost during this time, and William could see this was not a place he wanted to take up permanent residence.

With very little money and few opportunities for work after the flood, William bought a bunch of bananas at the dock and walked uptown, selling them on the street one by one to make a little change. He sold bananas like one would peddle candy bars today until he had enough money to buy a boat ticket and move on. It was a good thing he did, because as bad as the hurricanes, floods, and fires were when he was in Galveston, the horrendous hurricane soon to come in 1900 could well have killed him. It

Sheep pen

Going West

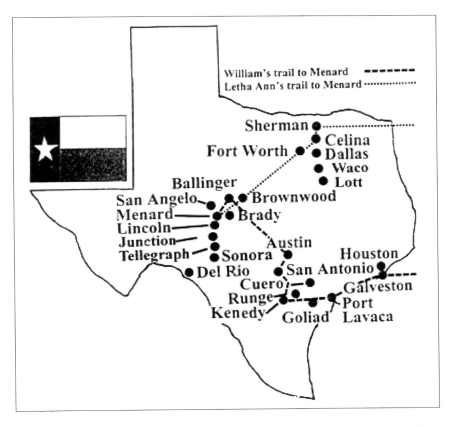

was the deadliest natural disaster ever to strike the United States, with at least eight thousand reported fatalities. So the good hand of God was upon William, guiding his steps once again.

Traveling on a Mallory steamship again, William went from Galveston to Port Lavaca, and then he walked a hundred miles to Victoria. At the time, he said that it was faster and easier to walk than to thumb a ride with one of the freighters. (Freighters in those days were large wagons pulled by teams of horses that moved products around the country much like tractor trailer trucks do today.) William later had his tool chest and a trunk shipped to him. From Victoria he moved to Karnes County near Kenedy, Texas, where, in 1877, knowing very little about sheep, he bought his first herd and then took more on shares from a man named Pat Rose. The ranges were open in those days, and there were very few fences in most of the country. Livestock rustling was rife, so deadly family feuds never seemed to end. William finally moved his herd close to the vicinity of his good friend Henry Hartgraves.

As he was building his herd, one year they lambed out five hundred lambs, and there was a lot of rainfall that year. William took the notion that the rains were helping to increase the range grass, and he believed he would finally, hopefully, be coming out ahead. He didn't know that a huge disaster lay just ahead. Shortly after this increase in moisture, his whole lamb crop began to develop stomach worms and die. William didn't quit, however, and go back to New York. He just studied the situation and then changed direction, going into breeding and trading stouter animals: horses and mules.

## Entering the Horse and Mule Business

William started breeding some of his own horses and mules as well as buying and selling the livestock. Most of his time was spent breaking broncs and harness training (or "gentling") mules. This, of course, made the animals more valuable to potential purchasers as well as making it possible for the animals to be transported by railroad in boxcars. During these days in Karnes County, he broke countless horses and had a stiff right ankle for the rest of his life to prove it.

Breaking a bronc in 1903

One horse in particular got to him a bit. William later said of this horse, "He was one of those that tries to fall over backwards" (on top of you). Once when this horse reared

William Menzies in the early 1880s

back, William stepped off, but his boot heel was worn and his ankle twisted when he hit the ground. He could hear and feel the ankle pop, leaving him in great pain, with no horse to ride and stranded about thirty miles out in the middle of nowhere. Fortunately, another cowboy soon showed up, and they tried their best to straighten William's ankle. The two of them bound it with strips of red flannel underwear. Then they returned to William's ranch, where he recuperated. It was more than six months before he could see a doctor about it, and at age ninety-one, he recalled, "I was already hobbling around then and there wasn't anything I could do."

Broncs were much bigger, heavier, and harder to break back then because most cowboys had to wait until the animals were four or five

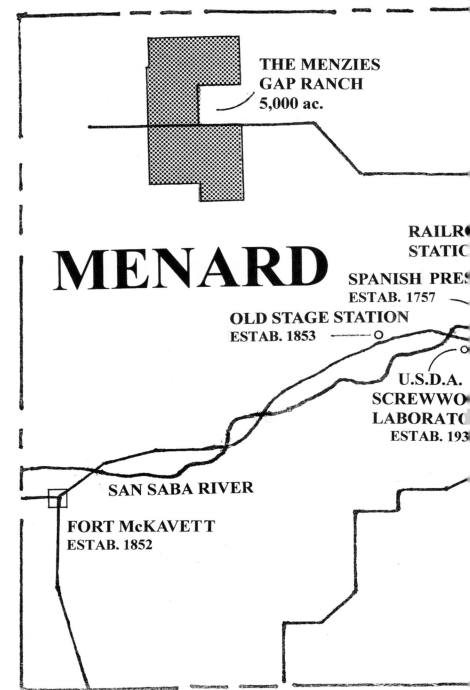

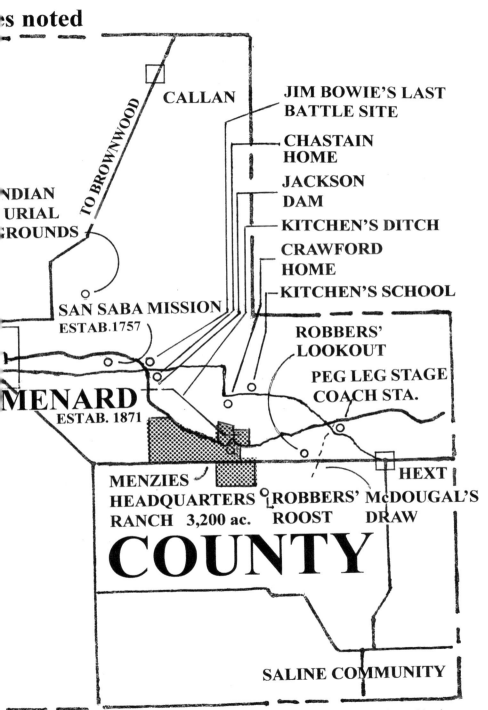

years old to do it. The reason behind this was that a horse's leg bones didn't mature until that age and could easily be broken. The cowboys' philosophy was to put them to work every day, all day long at the ranch or on a trail drive as soon as they were broke. Today, however, ranchers sometimes start breaking horses at two years old and start training them at three. But younger horses receive much better care nowadays, and they are not put into full service on ranches until they are four or five years old.

## Selling Horses Cross Country

After William recovered from his ankle injury, he decided to take a boxcar of horses to sell in New York, but nothing went easy for him. A strike on the Big Four Railroad caused him to take a southern route, driving the horses to Cuero, Texas, just to ship them by train. This also meant that he had to ship them through Mississippi and Georgia to get them to New York, where they were unloaded in Jersey City.

Roping a bronc to break

Rider, rope, and mid-1800s spur

During the ten years he lived in Karnes County, because of open ranging, a lot of animals strayed from their herds, which led to cattle rustling. This led to many deadly generational feuds. One of the most famous feuds involved the Butler, Elder, Barfield, and Sullivan families, which began like most other feuds—over cattle rustling. When William had to make a trip to Goliad, he returned by way of Buck Pedus's house. The Peduses were not only William's good friends but they were also generally good folks. William arrived after dark and shouted a hello before entering the house. The Peduses answered, "Come on in." But when he stepped through the door, two guns were jammed into his sides. The family wasn't taking any chances about being surprised by their enemies.

The territory was so dangerous at that time that one of William's friends did all his traveling after dark. He would get up at midnight and try to get to his destination before daybreak, because bushwhackers were a problem. These guys would hide in the bush and shoot at passersby as they rode down a road or trail. (Texas had more trails than roads at that time.) Most people aren't aware today that you can travel pretty well on horseback by moonlight if there aren't many clouds in the sky.

## William Goes West Again

William ranched for a while in Karnes County near Kenedy, Texas, but things didn't go well for him there for several reasons, not the least of which was the humid climate. Such humidity was not conducive to sheep and wool production. His friend Henry Hartgraves moved to beautiful Menardville, and William, feeling dissatisfied with his Karnes County operation, was ready for a change as well. Henry had told him that

Menard County was perfect for raising livestock, so he soon decided to move to Menardville as well. The name of the town was changed to Menard in 1911.

In July 1877 William arrived in Ballinger by train and then took the stagecoach from there to Menard. Fred Wilson, an early Menard pioneer, was the stagecoach driver that fateful, hot, dusty July day. Sullivan Brothers Bankers in San Antonio had given William a letter of introduction to Jim Callan, the county clerk. Jim was also a major property owner and a large stockholder in a bank in Menard. When William arrived in Menard, John Alex Smith took him out to look at a property known as the Eckart place, but William thought it was too isolated and too hard to find. Plus, there was no water source on the property. They spent the night at the Winslow property and rode down the beautiful San Saba River the next day. William decided to buy the Perry McConnel, Wallace, and Green Huey place, which consisted of two sections of land that had a lot of river frontage. The price was about $2.50 an acre. (A section is one square mile in area, or 640 acres.) The river was the deciding factor at the time and proved again and again that this purchase was a very wise decision.

With his purchase completed, William returned to Karnes County and drove his horses across country. At the time the Runge ranch had just fenced all of the Elm Country and Los Moras Country, throwing out all of the stray stock. The strays drifted down to the river for the water and ate up every bit of the grass on William's new property. Texas experienced a great drought from 1887 to 1889, and most creeks and rivers dried up into a few mudholes. Because of this, William had to pasture his horses with Jim Callan for the first winter. Ranchers like J. D. Landers near Fort McKavett drove their herds of horses and cattle north into Indian territory. This, though dangerous, allowed them to stay in the business until the life-giving rains returned.

Horses were in short supply because the Indians were buying as many horses as they could and stealing the rest if they had an opportunity. Gentled horses and mules were in great demand at that time. Recollecting about those days, William said that the Indians "would even come up to the cabin" (his first homeplace on the river) and "cut loose" the horses. (We would say they stole them.) Horses were so valuable and in such short supply at the time, William said, "One fellow who had a pretty good saddle pony claimed he was offered eighty acres for the horse, but he refused it."

Some of the leading men William dealt with in Menard County back then were Jim Callan (the county tax assessor who also had some banking interests), Dick Russel (the sheriff), and Henry Vanderstucken (the postmaster). Mr. Gay, the county clerk, later sold William a marriage license on December 19, 1888.

Just prior to William's arrival in Menard County, the people who lived east of Menardville, in the San Saba River area, were Mr. and Mrs. Arnett, Mr. and Mrs. Garrison, Mr. and Mrs. John Jackson, Mr. and Mrs. Holland, Mr. and Mrs. Bill Miller, Mrs. Morgan, Mr. C. P. Munley, the teacher and his family, as well as a Baptist preacher named Jolly. The leading merchant was the Paddy and Tull Smith family until Mr. Smith was killed by Indians in the 1870s. Then all of these people moved away.

During the 1870s a new wave of folks moved in, including the Adams, Chamberlains, Howells, and Sellers families. Living in this area was so tough and treacherous that after a while they all left as well. In the late 1870s came Oliver Russell and F. M. Kitchens. Then, in the 1880s, came Mr. and Mrs. L. M. Chastain and their family, the family of William Menzies's bride. As was previously mentioned, William came to this area in 1877.

## Chapter 3

# Settling in Menard County

**W**ILLIAM HAD NO PLACE in which to live on his newly purchased property, so he made arrangements with Jim and Clara Chastain, who were just across the river from his place. They accommodated him until he could build his house. The Chastains had thirteen children, so the living quarters were tight. They were certainly kind and neighborly enough to welcome William to stay with them, acting out of the Texas spirit, but little did any of them know that the friendship they forged then would blossom into generational bonds between the two families that have lasted to the present.

The Chastains' ranch

Ranchers in west Texas are always happy to be there for their neighbors in times of need. They think of it as an honor to be of service. After all, you might not find anyone else to help you for miles, and most of the time, the land is hot, dry, and hostile. And it is good not to forget that the day might come when you will be the one needing your neighbor's help.

William built his first homeplace on a flat knoll just southwest of a place that everyone in the Menzies family affectionately calls Grassy Point. This is a wooded place where the beautiful San Saba River winds and turns dramatically in such a U shape, or horseshoe, that it almost cuts back into itself. In this area the waters of the San Saba cleanse its clean, clear, cool self again and again over numerous rocky shoals, making a good number of pools, rapids, and ponds. The bed and most of the riverbank are covered with beautiful, small, smooth, rounded white rocks. The banks are covered with the lazy, shady greenery of burr oaks, pecan trees, sycamores, and elms as the river gently courses through the countryside. Cutting into its own banks, the river has made numerous fishing and swimming holes over the decades for several generations of kids to enjoy with delight.

Old bucket on a fence post

The first house William built was called a plank house. This kind of building was quite prestigious around Menard in the early pioneer days. It was constructed of heavy, rough-sawn beams and two-by-four walls. The exterior siding consisted of boards cut in one-by-twelve-inch planks nailed side by side with one-by-two-inch strips of wood (called bats) that were affixed to the walls to cover the cracks where the boards butted up against each other. Today, we call this siding technique or exterior treatment "boards and batten." There were precious few carpenters on the frontier, and most houses in Menard County were either log cabins or made by stacking rocks for walls and adding a wood pole roof system. Portland cement was hard to come by and next to nonexistent on the frontier. Having a plank house was quite expensive because all the lumber had to be hauled by horse-drawn wagons from Brownwood, which was more than a hundred miles away. There were no powered sawmills either locally or in any of the nearby cities, not even in Brownwood. Lumber had to be sawed in towns and cities farther east. Then it had to be shipped by rail to Brownwood. This made construction work on ranches extremely expensive and difficult because of the logistics.

Regardless of the high cost of constructing his home, the determined young William paid the price for progress as a pioneer. He put the construction skills he had learned from his father and in business in New York to use again and again throughout his life. His first house was actually two houses ten feet apart. Each building had one room and a porch on the front. The kitchen was a separate building behind the two main buildings.

## True Love's Beginning

The story of William and Letha Ann is just simply a love story. They met deep in the heart of Texas and loved each other just as deeply. They shared a love of God, their families, their children, their land, their work, their animals, their machinery, their community, and their country with this same kind of undying love.

While William was working on his ranch and building his new home, he continued to live with Jim Chastain's family across the river. Despite the close quarters for William, the couple, and their thirteen children, there were many happy times. Soon William met Jim's sister, Letha Ann, who was living with her parents on the ranch just west of them, near Five Mile Crossing. William started sparkin' with Letha Ann right off, and the romance was on.

Dances were quite popular at the time. After William began seeing Letha Ann, they enjoyed church services and dances a great deal. On one occasion, William, Letha Ann, Charlie Noyes, and Lula Kitchens (later Charlie's wife) all rode horses to the Fourth of July picnic and dance in Menard. Years later, Letha Ann and Lula remembered the new white muslin dresses they had made for the occasion. But by the time they arrived after a seven-mile ride in the Texas sun that July day, the girls' dresses were not as fresh and pretty as they had liked. Even so, they didn't worry a whole lot about it, because they danced for most of the night. Back then, ladies rode sidesaddle, and Letha Ann was quite an accomplished rider.

At the age of thirty-three, William had fallen in love with Letha Ann, and he was ready to tie the knot. After keeping company for about a year, the two were married by Judge J. D. Scruggs in 1888. Letha Ann was twenty-one. The judge came out from town to do the wedding at the Chastain place with only the immediate family and the Kitchens family in attendance. After the wedding and brief festivities, William and Letha Ann got in his old iron-axle wagon, rode across the San Saba River to the homeplace by the river, and set up housekeeping. They had no idea what was ahead of them down the road, but they had hope in God as big as Texas and just about as much determination in their hearts. Letha Ann wasted no time in acquiring a few Jersey milk cows and some chickens and planting her garden.

Theirs was a marriage made in heaven. Although they were quite different, they were a well-balanced pair. William was the enterprising, adventurous, risk-taking type. After all, what kind of a man would want to live his life where there was no risk or daring required? Not foolish risk taking for sure, but little can ever be accomplished in life if someone is afraid to take any risks at all. William also had a lot of common sense. He knew that sometimes you get the bear and sometimes the bear gets you, but you have to get up and keep going. What doesn't kill you will make you stronger—in time.

Letha Ann had a steady hand, contributing a conservative, level-headed approach in their decision making. She also had deep spiritual roots.

William was not a large man, just five feet nine inches tall, but he was trim, active, and smart. He loved horses, horse racing, building, machinery, and anything new or progressive. He had a huge zest for life. Letha Ann considered herself well favored to marry a very progressive New York Scotsman, since virtually no one had a college education on the frontier. Land-grant colleges did not come into existence until 1886 (other than one in Georgia in 1885), and then, at their beginnings, they were only fledgling institutions at best. William had advantages and skills with machinery, horse trading, and building contracting way beyond most young men of the West.

Living on their homestead on the river, William and Letha Ann made a success of the ranching operation for fifty-six years against all odds in very trying and difficult times, fighting bushwhackers, cattle rustlers, wolves, coyotes, boll weevils, droughts, disease, and floods, just to name a

William, Letha Ann, and George, c. 1890

few. The couple sent two sons, some grandsons, and great-grandsons to two world wars and lived through the Great Depression as well. Because they were willing to pay the price of perseverance, they saw progress, God's bounty, and enjoyed a great many good times.

## William's Brothers

When William came to Texas, his two brothers, Alex and George, stayed in New York. After he married and was fairly well settled on his ranch in

Alex and George Menzies

Menard, they both followed him there. Alex and George were initially plumbers by trade and, like William, both fancied horses and horse racing. However, neither of them saw this wild frontier as a place for them to live on a permanent basis. They had already established themselves in New York, assisting in their father's building business and in other commercial interests, so they returned very soon to the big city. Their stay at least gave them a love for the Texas hill country and a desire to make investments in land there, as well as to come back frequently to hunt and visit the family. George amassed a considerable interest in real estate and mortgages prior to his early death in 1912 at the age of fifty-five. Alex gained a sizable interest in a bank at Irvington, New York, and was always trading stocks on the stock exchange. We don't know the extent of Alex's holdings, because he died without an heir, but in his many letters he wrote to William about "fortunes being made and lost in a day" on the stock market. All three brothers maintained a close friendship throughout their lives.

Alex and George enjoyed visiting the family in Texas regularly to go hunting with William, although George did not visit as frequently as Alex. Alex obtained a seat on the New York stock exchange later in life and became very successful financially. He died in New York in 1936 at the age of seventy-seven.

## Horse Trading by Boxcar Loads

In 1897, about ten years after William and Letha Ann married, William took his second load of horses to New York and then shipped another load to

Oklahoma. The word at the time was that the federal government was going to give the Indians two million dollars to buy horses, and he made the sale in anticipation of that event. Communications were not the greatest in west Texas back then. The telegraph had been invented but only in several unusable forms in the early 1800s. The first electric telegraph company wasn't formed until 1845, and it did not come into wide usage until the 1860s and 1870s. It took even longer to get the lines out to Menard. Even when the telegraph was in optimal operation, this kind of communication was both crude and cumbersome. Anybody knows you can't make a deal selling horses using the telegraph and Morse code. The telephone, which is a great asset to commerce, wasn't invented in a practical form by Alexander Graham Bell until 1877, and it didn't come into anything close to wide usage across the country until the 1880s. Horse trading in west Texas back then was pretty much a word-of-mouth business.

Menard County and west Texas were still much the hinterland well into the 1890s. Even so, immediately after William made a deal on the horses, he drove them a hundred miles north to the Brownwood railhead and shipped them. His banking was handled by Goggins, Ford,

Just deciding who's boss

and Martin, bankers in Brownwood. First, ranchers had to drive their livestock to Brownwood and load them on the train there to ship to points beyond.

Tragically, a bank panic occurred at the very time William made the shipment, so when he went by the Brownwood bank where he did all his business, all he could get from them was ten dollars. While most people today do not know it, there were fifteen economic depressions during the nineteenth century—that is, depressions with a small d, more like severe recessions. The man who bought William's horses experienced the same problems in his area, and he couldn't get any money from his bank either. (There was no Federal Deposit Insurance Corporation back then. If there was a financial scare, the people would swarm the banks in droves and withdraw their money.)

William's buyer was finally able to borrow enough money from his friends, but he could only pay the freight bill. Then the purchaser's son rounded up two carloads of cattle and shipped them to St. Louis, where they were to be sold to get the money to pay for the horses. William went out with the man and helped him round up the cattle for the shipment. Then he had to wait there and look after his horses until the cows were sold in St. Louis. When the money was wired back to the boy's father, a financial settlement was made, and William was finally able to leave for home.

Breeding and selling horses across country was not the simplest business in an era with no trucks, phones or cell phones, Internet, or credit cards. For folks to stay in the horse-breeding business very long, selling them by the boxcar, they had to be gifted with a few natural talents. First, one had to be a skillful negotiator as well as being fleet of foot, a good judge of horseflesh, a good broncobuster, and a bit of a riverboat gambler. At the same time, you had to be circumspect, cautious, resourceful, and alert. Losing a sale on a boxcar load of horses to a deadbeat buyer several states away could have sunk most horse traders, and it would have put a huge dent in William's finances as well.

## The Spanish-American War

As the nineteenth century was winding down, things were going well on the Menzies ranch. And then in 1898 the Spanish-American War erupted over the liberation of Cuba and Spain's last colonies: the Philippine Islands, Puerto Rico, and Guam. Without a doubt, the most famous battle in this war was Teddy Roosevelt's Rough Riders charging up San Juan Hill. It was a short war, lasting from April to August 1898. But William supplied a good number of horses and mules to the government during this time. Of course, the war ended victoriously for the United States. Cuba was liberated, and the United States purchased the Philippines, Puerto Rico, and Guam for $20 million.

In addition to selling horses and mules, William pursued his passion: breeding racehorses. His horses raced in fairs and at get-togethers all over the country. He called his strain of horses a "steel dust breed." For years he kept about 150 broodmares and later started raising Morgan horses and mules. He stuck with raising horses and mules until about 1907, when it was clear the nation was turning quickly for its transportation needs to the horseless carriage.

## The Flood of 1899

In 1899, William took another of his many hundred-mile trips from home to buy lumber and other building materials in Brownwood. On such long trips he would harness three wild horses with one gentle horse to make his team and then tie a young mule or two to the side of these two teams. This was not at all easy to put together initially, but this is how he got the animals used to a harness and being yoked with others. The hundred-mile trek each way gave this training plenty of time to take place for the wild ones. But the trips never started out well.

Those mules and broncs would be kicking, snorting, and raising so much sand in the beginning that Letha Ann was afraid William would never get back—but he always did. Gentling horses and mules was just a part of what life was about for William. It was just a fact of life back then: every mule had to be gentled and every bronc had to be broke in order to sell and transport it.

There were still no sawmills in Menard or in the surrounding towns of Brady, Mason, or San Angelo, so sawn lumber had to bought in Brownwood and then brought a hundred miles by wagon all the way back to the ranch. William and Letha Ann were well into remodeling their home for an ever-increasing number of little ones, but unbeknownst to them, a terrible storm was swiftly converging on Menard County. It would be categorized as a hundred-year flood. And on June 9, 1899, the great flood was upon them. William was making headway on a lumber procurement trip, returning to the ranch from Brownwood when the rain began. He had to fight his way through rain, mud, and raging streams with a heavy wagon-load of lumber toward the north side of the river. The San Saba had begun to swell at about eight o'clock that morning, and it continued to rise into the night. It was so bad that William couldn't even make it as far back as the Chastain place. The river was already above flood stage. They didn't know it at the ranch, of course, but the floodwaters had already risen so high in the city of Menard, ten miles upstream, and nearly every house in town had been flooded.

Menardville on December 24, 1898.

Menardville during the flood of 1899.

*Settling in Menard County*

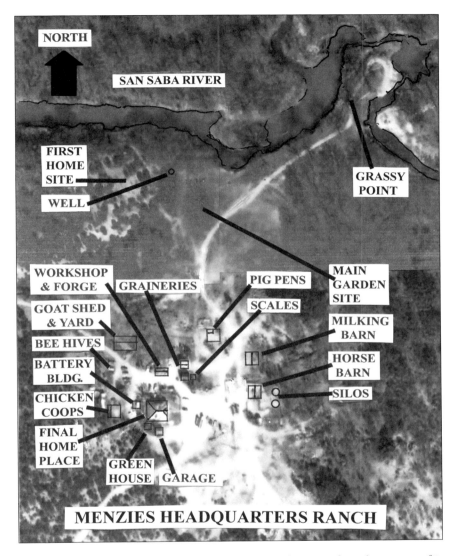

MENZIES HEADQUARTERS RANCH

    The raging flood was bad enough during daytime, but the worst of it came during the night. Letha Ann awoke and watched the river rise. She was quite alarmed, and when the chickens started squawking, she took a lantern and braved the rising water to rescue most of them. The water continued to rise and reached the top step of the front porch. She then decided to get everyone out. After gathering a few quilts, the five small kids, and the hired hand sick with chills and a fever, she made a break for it. She had the smallest child, Alex, in her arms, and the other kids cling-

Willie Lee Walston and a box of puppies at Granddad's forge

ing to her skirt. The water was upon them so fast that it was already ankle deep in the house as the kids clambered out of bed and put their feet on the floor. When Letha Ann stepped out the back door, she found herself in swiftly moving, waist-deep water. Singlehandedly, she took them all up the hill and into the saddle shed, out of harm's way, and there they spent the rest of the night. All night long the rain continued to fall and the river rose more and more.

Meanwhile, William was having his own troubles as he navigated home through the torrential rain. His loaded wagon continually bogged down on the roadway and in creeks that were also flooding. So he decided to abandon the wagon and the lumber. His first concern was his young family, so he loosed one of the team of horses and rode bareback, splashing down the road and swimming the rising creeks. Finally he reached the Chastain place, but he didn't know if his family was okay or not. His next obstacles were the two riverbeds running parallel to each other a good distance up the main river above the homeplace at Grassy Point. The second, slightly higher riverbed to the north was raging just as furiously as the southern riverbed and had swollen to the point that both riverbanks had merged into a single torrent of water filled with fast-moving logs and debris.

On the second day he fought his way through the river to get as close as he could to the homeplace. With the house in sight, he gave a shout and kept on calling out until he heard Letha Ann shout back. When he was sure she and the children were all right, he was greatly relieved and returned to the Chastain place to spend the night. Knowing it would be a few days before the river subsided enough for him to get across, but not knowing what would be left of his home, he decided to return to Brownwood for another load of lumber—to rebuild what he thought had been washed away. So he unloaded the lumber he had brought up from Brownwood and rode the hundred miles back to the lumber supply store.

About four feet of water flooded the old home, and everything in it was damaged. All the furniture fell apart and had to be glued back together. In a time before electricity and hydraulic pumps, people built their homes as close as they could to a water source because they had to haul water home by bucket. After this flood, William moved the house to the crest of the hill above Grassy Point, to the beautiful spot where it now stands as a quiet testimony to his undaunted will, industry, construction skills, and innate aesthetics. There his construction skills produced a number of barns, fences, silos, windmills, and outbuildings. William added his own forge and tool shop, where he spent innumerable hours building, designing, repairing, shoeing, and blacksmithing.

He moved the house uphill with just horses and mules. Having moved several houses myself, I can say that it is not easy to move a house today, even with heavy-duty trucks and hydraulic equipment. Moving a house with only horse and mule teams in William's day was even more difficult.

Much of William's time in the early years of his marriage was spent hauling lumber from Brownwood to the homeplace for fences, barns, and remodeling his home to take care of his growing family. Building fences, houses, and barns requires a tremendous volume of material and hardware. Add the transportation factor to this, and subtract trucks and free deliveries from local lumber suppliers, and you're looking at long, arduous hours on bumpy, crooked dirt roads in the hot summer sun or bitter wintry winds. The hundred-mile road also made you easy prey for numerous roadside bushwhackers. If you were headed to market and your wagon was empty, the bandits knew you probably had money on you to buy something, and if you were headed home with supplies, they knew they could take all that you had by knocking you off with a single rifle shot.

When Walter Menzies (William's grandson) was in his early teens, he accompanied William on one of these supply and lumber runs by wagon to Ballinger. On the way back, as the sun was setting, William pulled off the road, set up camp, and fixed their supper. After they had eaten, three very unsavory men rode up. After their conversation went on a little too long, William sensed that the men had ulterior motives. He pulled out the pistol he always kept in his belt and laid it on his bedroll. He then told the men that Walter and he had a long way to go the next day and that they needed to sleep, suggesting that the men needed to leave the camp and ride off. As soon as the men were out of sight, William told Walter to break camp and throw their bedrolls on the wagon. They ran the horses down the road a good distance in the dark before making camp again. But this time they set up their camp two or three miles off the road, where they could hide their wagon in the brush. They slept there the rest of the night without incident and drove home the next day.

## Entering the Cattle Business

William purchased his first cattle with his profits from trading horses and mules. He gradually worked his way into the cattle business, raising high-grade cattle for a number of years. His first two registered Hereford cows

Hereford bull

were bought from a man in Williamson County. Much later, in 1923, he bought forty heifers from the Bunger herd at Eden, and from them he was able to build up his herd by buying bulls and individual cows with the breeding he sought. Being successful in the ranching business required an alert eye to the constantly changing market, an ability to grow your own feed, and having just the right mix of the kind of animals that happened to be in demand from time to time.

## Chapter 4

# Letha Ann

Letha Ann, known by many affectionately as Lee, was born on October 2, 1867, in Collin County, Texas. Her father, Littleton Maxwell "L. M." Chastain, was a native of South Carolina and had lived in Georgia before coming to Texas.

According to the Chastain family history, their clan was from France, which their forebears had fled because of religious persecution. The French Protestants, known as Huguenots, were a branch of Calvinists who were persecuted by French Catholics to the point of frequently taking up arms. This constant clashing steadily decreased their numbers. During the St. Bartholomew's Day massacre of 1572 somewhere between five thousand and thirty thousand Huguenots were killed, including many of the most wealthy and prominent Huguenots in the country who had come to Paris to witness a royal marriage. Later, at least two hundred thousand Huguenots

Mary and eight-year-old Letha Ann

were driven from France by a series of persecutions. They fled to the Netherlands, Switzerland, Italy, England, and the New World.

The earliest ancestor whom the family has traced on Letha Ann's side was Dr. Pierre Chastain, a French physician. He and his wife, Susanne Reynaud, fled to Switzerland and lived there for about ten years where they had eight children. However, only five of the children survived infancy.

Pierre assisted in gathering a group of his Huguenot brethren and negotiated a pact with some English merchants to help colonize Virginia in exchange for land grants and transportation to the Americas. The actual ten thousand acres in Virginia chosen for the Huguenots was just west of the James River. Sadly, as was the case with the original Jamestown settlers, the first winter took a heavy toll on these pilgrims. Susanne and three children were among those who died that first year. Only one of the Chastain children lived to adulthood.

About a year later, Pierre married another Huguenot, Anne Soblet, and they had eight children. One of their sons was Pierre Chastain Jr., who later married Mildred Archer and fathered nine children. One son from this marriage, born in 1743 in Goochland County, Virginia, was John Chastain. He became a Baptist minister and married an Irish immigrant, Mary O'Bryan. They had eleven children who reached adulthood. After his first wife died, John married Mary Robertson.

John had grown up in the Anglican Church, which was the only denomination that his county recognized as legitimate. But the Baptist doctrine he later embraced led to severe persecution, threatening his life and ministry. He then blazed a gospel trail to the south, leaving Virginia as well as the influence and persecution of the Anglican Church.

His ministry led him to the mountainous region west of Greenville, South Carolina, where he raised his family and ministered as a Baptist preacher. In 1795 he founded Oolenoy Baptist Church in Pumpkinville, South Carolina, and served as the congregation's first pastor.

During the Revolutionary War, John openly defied George III with his sermons. He was placed on the king's list of treasonous subjects and was scheduled to be flogged. He may have served as a Revolutionary soldier in Virginia, but it is known that he was a signer of the Oath of Allegiance in Powhatan, Virginia, during this critical and dangerous time. The Daughters of the American Revolution later designated him as a patriot for this as well as for his resisting the religious restrictions imposed by the Crown. He and his brother, James, traveled up and down the frontier, preaching the gospel,

Letha Ann Menzies

founding and shepherding a number of churches. The two of them helped ignite the fires of the early Baptist movement in the United States. John had such a clear and resounding voice that he was nicknamed "Ten-Shilling Bell." (Apparently the cost determined the size of the bell, and ten shillings would have purchased a pretty sizable one at the time.) In a day without electronic amplification, it was said that he could easily be heard for a mile.

John was also a member of the Black-Robed Regiment, a group of preachers during the Revolutionary War who were not afraid to speak out on the issues of the day. The British believed that this brigade of preachers kindled the fires of revolution and kept it aflame. They also believed these preachers were ultimately responsible for the colonists being able to win their independence. They were terribly lacking in munitions, training, clothes, shoes, and numbers. They believed God compensated for this lack by firing up the zeal of the colonists and the spirit of independence through the potent sermons of these firebrand preachers. American preachers have always had an anointing of the Spirit of God greater than or equal to the task before them. These fearless black-robed clergymen would proclaim the truth from their pulpits regardless of the consequences. They preached that the only way for America to be free was for its people to be independent from England and self-reliant but governed within by a new king. America, they said, could only be a great nation under the kingship of the Lord Jesus Christ.

L. M. Chastain, John's great-grandson and Letha Ann's father, came to Texas in 1858. But shortly after that, he served in the Civil War as a Confederate infantryman. After the war he returned to Texas via South Carolina and Georgia and settled for a while in Grayson County, Texas. He later moved his family to Collin County, north of Dallas, and became involved in farming and raising livestock.

Letha Ann's mother, Alcy Alexander, was also a native of South Carolina. She was born on August 16, 1833, and married L. M. on December 20, 1849, when she was sixteen years old. The Chastains had seven children: Jim, Letha Ann, Evaline, Maxwell, Mary, William, and Perry. L. M. was a very successful farmer and rancher and was able to increase his herds considerably over the years.

In 1877, when Letha Ann was ten, the family moved to Menard County and bought a section or more of land just east of Menard, below Five Mile Crossing along the San Saba River. Their home was on the road leading to what would later be the Menzies ranch. Letha Ann didn't know then that down that same road, eleven years later, a dashing young horse trader from south Texas would ride up to their house. She didn't know this young cowboy would win her heart, sweep her off of her feet, and keep her busy with the joys of raising a large family and building ranches for the rest of her life.

Her father had originally scouted the San Saba River area before the Civil War. He knew the land along the river was beautiful country, and he never forgot it. The Chastains came to Menard during the summer to purchase this land and then return to Collin County to fetch their livestock and belongings. They made the trip back to Menard with three other families, including the Perry McConnel family and the Dave Alexander family. In all they had six covered wagons and quite a large herd of cattle. The cattle slowed them down, straying from the bed-ground and doing what cattle always do. They also had to leave some of the calves behind. Quite frequently they had to circle back several miles after losing some livestock. It took them six weeks to reach their new home.

The Chastains' first home was a double two-story log cabin. It had two rooms upstairs and two rooms down, with a kitchen and dining room built on. They also had a workroom, a cellar, several barns, and a cistern that was filtered by a box of gravel and charcoal. Letha Ann lived there with her family, helping with the work and sharing in the pleasures of their day. Most of their recreation was going to church at the old Kitchens schoolhouse and attending camp meetings during the summer.

Sometimes at Christmas Letha Ann would visit her brother Max's home near town, to spend the night, carrying her party dress wrapped up in her shawl. Then they would ride to Menard to the Christmas dance held at the old schoolhouse. These dances were so popular that, as the town began to grow, a dance hall was soon built.

## The Itinerate Preacher

At about the time that William began to court Letha Ann, an itinerate preacher arrived for a brief stay at the Chastain home. During his late October visit, the whirling fury of a Texas norther descended on the county. Preachers were quite poor back then, and this one was no exception; he owned only the clothes he wore. Since eleven people were living in the house at the time, he wanted to clean his suit—and improve his own hygiene as a courtesy to the others. He headed down to the river and dutifully removed and washed his clothes. It was already quite cold, but as the norther blew in, a hard freeze fell. Much to the preacher's surprise, his clothes were frozen stiff in just a minute or two as he tried to dry them. They froze so hard he couldn't get all of them back on. As a result of the exposure, he came down with pneumonia and died not long afterward. Letha Ann's parents sent someone to find William so he could help with the body and give the preacher a proper burial. Such was the harshness of frontier life.

## A Pioneer Wife

In Texas in the nineteenth century, a man needed a sturdy, virtuous, God-fearing wife. Letha Ann was all that and more *and* beautiful besides.

Woodburning stove

She had a peaches-and-cream complexion and fine teeth that she kept her whole life. Her eyes were blue, her hair light brown, and she was about five feet eight inches tall. Most important, if she was nothing else, Letha Ann was a faithful Proverbs 31 woman. A meek and quiet spirit enhanced her beauty. In short, she was generous, godly, and gracious. In addition to birthing, raising, and tending to eight children, she also worked hard on the ranch. Not only was she a homemaker who fixed three meals a day for ten people, but she also did the washing, mending and making clothes, baking bread, and helping milk cows in the dairy.

If that wasn't enough, Letha Ann was also a keen businesswoman. Just after William and she were married, she had some chickens, a few Jersey cows, and a large garden from which she began to supply Menardville with country produce. After the Kitchens Irrigation Canal was organized and constructed, Letha Ann enlarged her garden. They had all kinds of vegetables, including green beans, pinto beans, tomatoes, sweet potatoes, pumpkins, corn, cushaw, and watermelons. In their orchard, the Menzies had peaches, pears, pecans, and figs. They tried to have a tomato canning plant for a while and actually had their own label. But it failed to catch on in the community and soon closed. This garden—along with milk, butter,

Alex and Letha Ann.

cottage cheese, chickens, fish, turkeys, lamb, hogs, beef, and honey—supplied a bountiful table for this growing family. They also had a cool,

Above, fishing on the San Saba. Below, Bill Menzies (left) and George Faber (right) take a short break from fishing.

dry cellar under the house where Letha Ann stored their canned goods, sweet potatoes, dried beans, and onions.

Letha Ann's most outstanding characteristic was her love for people. Letters from William's mother, Agnes, are evidence of a regular correspondence. Agnes referred to her daughter-in-law as "my darling daughter." The letters indicated that Agnes sent the small children, George and Agnes, ten dollars each as birthday presents. That was a huge amount of money back then. She also made

Washday equipment

A picnic at Grassy Point. From left to right, Mary Pearl McWilliams, Perry Menzies, Willie Lee Walston, Billie Anne Menzies in bonnet, Ray Jacoby, Letha Ann Menzies, Raymond Roy Walston, Roy Jacoby, Agnes Walston with hat, and Pearl McWilliams.

and shipped many clothes to Letha Ann, including her maternity dresses. Letha Ann needed those quite frequently with all the kids she was having. She showed a deep interest in William's mother and looked after her own mother when the time came. She was devoted to her brothers and sisters as well. After the death of her eldest son George in 1920, she took time to visit her family in Pilot Point, Fort Worth, and Denison every few years. The Chastain family, in turn, would visit the Menard homeplace nearly every summer. And everyone had a wonderful time during these visits.

The extended family was always welcome, loved, well fed, and refreshed after a summer vacation at the Menzies ranch on the San Saba. Many family picnics occurred at Grassy Point on the river.

Letha Ann's hospitality and giving spirit were known far and wide by their children, their neighbors and relatives, their friends, visitors, their preacher on Sunday, travelers, and even "down and outers" who came by. A traveling salesman named Chidnster regularly stopped by the Menzies ranch, ostensibly to sell insurance. Letha Ann said he could always smell when there was a fresh churn of buttermilk, because he always appeared soon after. She began to call him Buttermilk Chidnster. Her welcome greeting to visitors has since been handed down: "Come in and eat. We

Letha Ann holds a baby with Pearl and her turkeys.

Dairy and cooking implements

have already eaten, but there is plenty of it, such as it is, and good enough, what there is of it."

On one occasion, Rusty Williamson rode with his dad out to the ranch to buy a Hereford bull from William. While there, he saw a pair of fox terrier puppies and took a shine to one of them, so Letha Ann gave it to him. Rusty loved that little dog, but tragically, his dad was having a problem with packs of wild dogs killing sheep on their ranch in the southeastern part of Menard County. He put out some strychnine poison to kill the wild dogs, but Rusty's little dog accidentally got into it and died. When Letha Ann heard about it, she gave Rusty a dog from a different litter. Rusty loved his new dog more than the first one. He said that his dog "was a varmint getter and one of the smartest little dogs I ever saw." When Rusty went fox hunting with his hounds, he would lean down while on horseback and grab that little dog by the collar. As soon as the dog felt Rusty's hand on his collar, he knew to leap with his back legs while Rusty pulled him up on the horse with him. Sitting behind the saddle on the horse's rump, the dog balanced himself there and never fell off, no matter what they got into or how the horse turned, dodged, or leaped.

With eight kids, Letha Ann had to cook for ten people three times a day, but there were rarely just ten people at the house. The Menzies home was the meeting place for the ranching community east of town. One or two of the kids always had a friend over. My dad and his brother were there a lot, along with all kinds of other "in-laws and out-laws," as they say. A great deal of wood also had to be hauled and chopped for Letha Ann's wood-burning stove.

Letha Ann also helped William with his beehives, as well as tending to her own animals. She raised chickens, hogs, turkeys, geese, ducks, sheep, lambs, and guineas. She even made her own down pillows, which were considered a special blessing at that time. Letha Ann was the first to raise sheep in Menard, and then she got William interested in them again. Sheep were very productive animals to have in this predominately arid country. They could find and eat vegetation that horses and cows either weren't able to get to or weren't interested in.

## William's Dairy

Butter churn

Before World War I, in addition to raising children, constructing, farming, gardening, operating a mobile food sales operation, and ranching, William and Letha Ann also found time to start a dairy. William bought a boxcar of short horns and Jersey cows, and from that he nearly always kept a twenty-cow dairy. From that point on, it took four or five people a day, including Letha Ann, when she could, to do the milking. And cows don't take weekends and holidays off. They had to be rounded up, put in stalls, fed, cared for, and milked twice a day, early in the morning and again in the evening. Most cows give an average of eight gallons of milk a day, and some may give as high as sixteen. Back then, they had to sit on a stool and milk the cows. With twenty cows, the family had to deal

The family hack in front of the homeplace, with Alex on the left and Bill on the right.

Letha's hack at the Luckenbach Hardware store.

with at least 160 gallons of milk a day. But the work didn't stop there. The family also had a steel separator tank, and they churned twenty gallons of cream at a time. The churn was a wooden barrel with side braces and a short wheel so they could shake the cream into butter. After all the work was out of the way, the dairy proved to be a great blessing to the family. They did all of this without the benefit of electricity, refrigeration, homogenization, pasteurization, plastic, or modern containerization. If either William or Letha Ann had a lazy bone in their bodies, they never found it. They knew only too well that the only place where *success* comes before *work* is in the dictionary.

Letha Ann grew all her crops, raised her own animals, and did it all without the assistance of electrical appliances and other conveniences that are taken for granted today. But it doesn't stop there. Letha Ann also had to market much of what she produced, supplying produce to the residents of Menard. She had her own home-delivery distribution company and her own checking account, which was most uncommon in her day. On Saturdays, with the help of some of the children, she loaded her wagon with a variety of food, such as five bushels of tomatoes, a hundred pounds of butter, several half gallons of buttermilk, some honey, some chickens, dozens of eggs, and other salable items: peaches, figs, and pecans. Then she would cross the river and drive her wagon eight miles to town. The kids would help her deliver the products to their steady customers around town, and they seldom had anything left over that had to be taken back home. She was able to supply the townspeople with a great many things that the local stores didn't offer.

In addition to this, Letha Ann had certain days that were "light bread" baking days. If all this wasn't enough, she made her own lye soap (as most did back then) in a huge iron kettle behind the house. She would put in all the ingredients, heat it up with a wood fire, and then let it cool. Once it solidified, she would cut it into blocks with a knife. All through her married life, she canned and preserved various foods. While she was doing all this, William was raising sheep, goats, pigs, cows, mules, and horses.

With her own checking account for the family business, she not only bought clothes for the family but also made a good many of them herself. Amazingly, she also provided most of the food for the burgeoning family. Her producing and paying for the basic necessities of life for the family allowed William to use his finances to buy more animals and land as well as to make needed improvements. He was also able to buy

A homemade Menzies quilt

the most efficient machinery and equipment of the day. Their ranch became an industry in itself.

By 1912 Letha Ann owned her own flock of sheep. With her profits from this and her food sales, she was able to buy some household furnishings. She bought a dining room set with an oval glass china cabinet, which was shipped from San Antonio. She purchased a piano, lace curtains, and some very pretty quilts as well as some utility quilts. Her homemade quilts were made from leftover fabric, cloth from feed sacks, as well as the good parts and pieces of old clothes. Recycling is not a new thing; it was invented a long time ago. Her quilts were works of art and love. They each required no fewer than a million stitches just for the ornamentation. Of course, she did this in her "free time."

Letha Ann was always proud of her home in those days. She had a yard and a porch that were full of flowers in the summer, and she stored potted plants in her concrete greenhouse during the winter.

William gave her one of his best pastures to raise her sheep. When he died and the property was divided among the kids, it was said that the east pasture at the old homeplace on the San Saba had always been "babied," since that was where Letha Ann ran her sheep. In addition, her sheep were bigger than any of the others being raised in the county.

## Home Remedies

Letha Ann formulated her own home remedies to heal the many ailments of the kids. Once, all the kids were working in the huge garden, hoeing weeds. To do that they had to keep their hoes sharp, because they didn't have any herbicides back then. As the kids were wont to do sometimes, they goofed off by playing stick horses and riding their hoes like wild horses. Accidentally, one of the hoes popped up and hit Alex in the ankle, cutting a tendon in half and leaving a severely bleeding wound. The other children took him straight to Letha Ann, who cleaned out the wound and then poured in a home potion and wrapped it with a cloth bandage. Then she poured turpentine over it and gave it a good season of prayer. Believe it or not, God healed Alex, and he never had a problem with his tendon—even while playing football in college.

Letha Ann was also ready to assist with delivering babies for the neighbors. At the age of forty-three, when her youngest child was eight years old, Letha Ann was pregnant again. William summoned Dr. McK-

Alex and Max Menzies with a calf

night to deliver Letha Ann's babies, but the doc didn't get there in time. The family says that Doc came out for a delivery and found Letha Ann in bed and the baby by her side.

Alex remembered a time when Max had osteomyelitis in his arm for a number of years. His arm would swell up occasionally, and Letha Ann would use Madeira vine leaves for a poultice. Once his arm was so swollen they took him to see Dr. Findley. Alex wanted to go along, too, because he had gotten all the leaves for the poultice and wanted to see the results of all this. When Doc pierced Max's arm, the infection shot straight out. Alex immediately fainted and fell flat on the floor. When he came to, he nearly fainted again because he was very bashful and found himself looking into the face of a beautiful nurse at close range.

Because of hard work, fresh air, good food, and a wholesome Christian lifestyle, the family was usually healthy most of the time. Once, though, William hurt his back, so Letha Ann wrapped a hot stove cap with paper and cloth and laid it on the affected area. It cured his backache. With God's help, Letha Ann could fix anything.

## William's Greatest Asset

Letha Ann was truly one of the greatest reasons for William's phenomenal success. As the old adage goes, "Behind every successful man is a great woman," and in William's case, this was never more true. Over fifty-six years of married life, their faithful love and devotion to one another and to God made their many successes and victories possible. At the same time, it would be an understatement to say they had many problems. One of the biggest problems the family faced was the Great Depression. From 1929 to 1932, unemployment in across America went from 3.2 percent to a high of 24 percent. In three short years, the economy shrank 43 percent from $103.6 billion to $58.7 billion and more than a thousand banks failed, costing many hardworking Americans nearly every penny they had. Out of this disaster came the Federal Deposit Insurance Corporation, which was just one of the bank reforms made as a result of the Great Depression.

The long arm of lack sprouted tentacles that reached into every neighborhood, town, and city across America, from the country's biggest cities to the most remote ranches of west Texas. During the Depression, when it was difficult for William and Letha Ann to buy shoes for their children, they said, "Because of Mother, we always ate good, no matter what."

Letha Ann held the family accountable to the high moral standard of the family name. Once she reprimanded one of the children by saying, "A Menzies wouldn't do that!" That could be the best advice she ever passed on to the future generations who bear the name.

On one occasion, William had left town by train on a horseracing trip to see his brothers in New York. The bankers in Menard had recently loaned him about five hundred dollars (a large sum back then). When they heard that he had left the state, they thought his leaving somehow meant that he had defaulted on the loan. The bankers came out to the ranch and told Letha Ann that they were calling in the loan. She could either pay it off in cash or they were going to take the tract of land that collateralized William's debt. We've already touched on Letha Ann's industry and country produce business. Since she had numerous customers who took advantage of her food delivery of eggs, honey, butter, meat, and produce, she immediately visited every customer on her route. In one day she raised or borrowed enough money to pay off the debt. Through her efforts alone, the bankers failed to foreclose on the land they were after.

When Letha Ann's dad, L. M., passed away in 1903, her mother, Alcy, continued to live in her own home and take care of the place as best she could under the watchful eye of Letha Ann. The children always remembered their grandmother because she not only made the greatest jam and jelly but she also could be counted on to have a big sandwich ready for them whenever they returned from trips to town. For a long time, as her health declined, the eight Chastain children took turns living with her for a week at a time at the ranch at Five Mile Crossing. When Alcy became bedridden, Letha Ann and William brought her into their home permanently. Each of the Menzies kids went to her room every morning to give her a kiss and bid her a good day before leaving the house for work or school. They cared for her until she died on January 11, 1913. She was buried alongside L. M. at the Pioneer Rest Cemetery in Menard.

## Letha Ann's Devotion

Letha Ann was deeply devoted to William. She always looked after William's health, happiness, and well-being. Every morning, instead of taking a coffee break, William would come up to the house from wherever he was on the ranch to see her and have a cold glass of buttermilk. With eight kids and a lot of hired hands, he needed time to be alone with her to talk and plan. Those times were sweet. She would also special order salted mackerel for him, because she knew he was fond of it. During her last illness, she was still tending to William's happiness from her hospital bed. Her last instruction was to send her son Bill to find some salted mackerel to take home to him. What devotion!

Certainly, in this family, as in all others, there were times of disappointment, sorrow, mistakes, friction, and tragedy. But Letha Ann had the spiritual and physical fortitude to cope with whatever life sent her way. At the time of Letha Ann's passing on January 1, 1945, Mrs. A. W. Noguess wrote this and submitted it to the *Menard News*:

> January first witnessed the passing of a great and good woman, Mrs. Letha Ann Chastain Menzies. She was a good daughter, a good wife and a good mother, also a good neighbor. When Max Menzies, her seventh child was three or four months old, there was a neighbor woman who died and left a little baby. Well, Mrs. Letha Ann Menzies came and took the little one to raise, she thought she could nurse both the children, but found she did not have enough milk for two babies and had to let the little Hardwick baby go to Alabama to one of the little one's aunts to be raised. But this noble woman was willing to make room in her home for the little, motherless baby, while she had seven little ones of her own to care for and a trip once a week to bring butter, milk, eggs and vegetables to Menard to her customers. Mr. A. E. Nauwald remarked not long ago that when he first moved to Menard they got all their supplies of that kind from Mrs. Menzies and he did not know what they would have done without them. She was a very useful woman. There are few women like Letha Ann Chastain Menzies.

## Chapter 5

# Character of These Pioneers

Letha Ann was wise because of her faith in God and her knowledge of His Word. My dad, Perry Menzies, was semiadopted by William and Letha Ann. When he was three years old, he lost his father, so he and his older brother, G. C., lived with their grandparents from time to time, especially during the summers. He often talked about her, recalling, "Grandmother was always quoting Proverbs and had so many wise sayings." She was a sage and a rock of faith for the whole family as well as a cornerstone of the church. The whole family was a major part and support of their church. They originally attended what was called Kitchens Baptist Church, which was organized on October 12, 1879, and had Z. M. Wells as the first pastor. They held their meetings at the Kitchens School just east of Five Mile Crossing for many years.

Kitchens School, Class of 1908

The family went to Sunday school and church just about every Sunday morning. On Sunday nights there would be church and singings. William and Letha Ann contributed to the church finances and events, including organizing brush arbor dinners in the summers. Their lives were a sterling example to their family and their neighbors of what can happen when a young couple humbles themselves under the mighty hand of God. It was no small task just to get everyone to church on Sunday, let alone on time. It meant a river crossing and a three-mile junket by horse and buggy. As to the problem of getting them to church on time, Letha Ann would say, "We are going to be late with church half over." It was quite a problem getting all the kids ready, so William's reply usually would be, "Come on, come on; half of it is in the going." He and Letha Ann would ride on the front seat of the family hack with Walter in the middle and the three girls riding on the seat behind them. The other boys—George, Bill, Alex, and Max—rode on horseback. They persisted in going to church even in the rainy, windy, hot, snowy, and icy weather, and they always thought it well worth the trouble for their family's benefit. Letha Ann was always faithful to her family and friends and was widely known for her stalwart, sterling character.

Kitchens Baptist Church grew, as did the Lord's work. However, in 1916, automobiles became prevalent in the county, and the congregation built a wooden structure at its present site in town, merging with Menard First Baptist Church. That building was subsequently razed and replaced with the current brick edifice in 1950.

The foundation of Menard County has just about always been its churches. The strength, stability, and character they infused in the community have truly been of inestimable value over the years. In 1980, just about every denomination had a strong presence in Menard, including Calvary Episcopal Church, Christian Church, First Presbyterian Church, Grace Lutheran Church, Menard Bible Baptist Church, Menard Church of Christ, Sacred Heart (Catholic) Church, Seventh-Day Adventist Church, South Side Church of Christ, St. James Episcopal Church, and United Methodist Church.

## A Rancher's Industry and Self-Reliance

Ranchers in west Texas lived far out in the wilds, mostly on large spreads, and knew very well that when a fence broke down, there was no sense

*Character of These Pioneers*

Old chute

crying about it, looking for a government program to fix it, or calling for someone in the business to replace it. They knew they had better get out and fix it themselves. If a truck broke down, they better not wait on the backside of three sections of land for a mechanic to wander by. They better know how to fix it themselves.

I will never forget the time in east Texas when I was about ten years old and we were visiting my maternal grandfather, Herman Friudenberg. My older brother Steve and I called him Pop. He had a feedstore and a house in Lott, Texas, which is just south of Waco. He also had a small ranch, farm, and house outside of town. At that time, Pop, Steve, and I were tearing down an old house, pulling out all the nails and saving the lumber to build an addition to Pop's home in town. At ten years old, this was my first shot at home demolition, and I remember thinking that tearing down houses could possibly be the hardest, hottest, and dirtiest job in the world. Now, some fifty years later and after having taken down several more houses since then, I still believe I might have been right about that. But it was well worth the trouble, because we were able to build three additions to Pop's house and had a tremendous amount of lumber left over, which we stored in the warehouse behind his feedstore. Pop had enough leftover lumber to do whatever he wanted in the way of construction for a long, long time.

Granddad Friudenberg and the author

Around that time I lifted the hood on his 1950 Ford truck for some reason and noticed he had repaired the exhaust pipe coming out of the manifold by wrapping it with a metal Birds Eye orange juice can and tying it off with baling wire. That's what you call an inexpensive remedy.

During a drought in the 1950s, Pop's well dried up at the ranch and the cattle had no water. He didn't quit. He put a bunch of fifty-five-gallon barrels on the back of his pickup truck and hauled water from the city reservoir on the other side of town just about every day or so for several years.

Both sides of my family in Texas have that same self-reliant, never-quit, stay-at-it–until-you-get-it, never-say-die spirit. Ranchers in those days had to be handy with tools, creative, and resourceful just to get by. They had to learn quickly how to save, conserve, and make do with what they had. When they killed a hog, the only thing they wasted was the squeal. Texans lived by the old adage, "Use it up, wear it out, make it do, or do without." Sometimes that wouldn't even do. They had to mend it, bend it, stretch it, modify it, or substitute it for something else. It was a good thing they were faithful in the small things, because fortune has always seemed to favor the efficient. They did, however, have one other secret weapon that gave them a huge, unfair advantage: baling wire (pronounced: "balin' war"). With a little baling wire, ranchers could fix just about anything from a broken heart to the crack of the break of day.

Self-reliance was also very important in the critical area of doctoring livestock. If the sheep came down with screwworms, a rancher knew he could not afford to call a veterinarian out to the place for $125 to tend a $25 animal. He knew he had better get at it, deworming and doctoring the animal himself. That is why Texas ranchers are a something-for-something and nothing-for-nothing kind of people. You couldn't find a more kind and generous bunch of folks anywhere. They just don't like the liberal giveaway programs that take from those who work hard and give to those who refuse to work at all. That's why they have historically been conservatives. Oh, they believe in helping the helpless, but they also believe in the Christian work ethic: "If anyone will not work, neither shall he eat" (2 Thessalonians 3:10). This country has become great because of a lot of hard work, faith, and self-reliance, not because of government giveaway programs.

## The Gap Ranch

In 1913, after William had the headquarters ranch east of Menard fenced, stocked, and improved, he acquired several sections of land north of town on Wilhelm Lane. He called it the Gap property. It may have been called by many other names in times past, but now we just call it the Gap. It is the only place in those parts where a couple of small rocky ridges run pretty close together, and the old roadbed runs right through them. Both ridges are covered with live oaks, mesquite trees, and all kinds of vegetation. The area between the two ridges is the Gap itself.

As soon as William bought the ranch, he fenced it and moved huge herds of sheep and cows out there from the headquarters ranch, which was east of Menard. There was no such thing as a tractor trailer truck or a cattle trailer back then. Sheep were herded on foot. Many times the older boys—Alex, Bill, and Max—had to herd the sheep out there on foot, and most of the time the boys were without any shoes. William wasn't trying to be hard on them. Times were tough, and they couldn't afford shoes. It is a long way from the headquarters ranch out to the Gap—twenty-three miles. Herding the sheep on foot usually took the boys two days to make the trip, one way. Was it easy? No. Was it hot in the west Texas summertime? Yes. Was it cold in the winter? Yes. Was it a lot of fun? No. Was anyone else going to do it for them? No. Could they call Austin or Washington for any kind of government assistance? No, and they wouldn't even have thought of trying to do so. Did it have to be done? Yes. The boys did it time after time after time when they were in their early teens. They always got the job done.

## Chapter 6

# Ranching in West Texas

Ranching successfully on arid west Texas land takes the right mix and number of animals. Foraging animals harvest this land's natural foliage without damaging nature's ability to produce more. Overstocking or improper stocking can damage the ecosystem of a ranch for decades. In addition to that, a rancher has to keep his eye on the constantly changing demands of the market in his area.

William's first venture in Texas was raising sheep in Karnes County, but that didn't work out very well. Because of the area's heavy rainfall and humidity, the sheep became "wormy." That experience was a bitter pill for

Delaine sheep

him to swallow, and he learned that several years of hard work and investment could be wiped out overnight in the ranching business. One's choice of animals and proper stocking were crucial factors for long-term success. Letha Ann managed to rekindle his interest in sheep. Part of the reason William was reluctant to go back into sheep raising was his concern with the impact of coyotes and wolves on the sheep in west Texas. However, those reservations were overcome. There were two main considerations: Menard County is arid country, making it excellent sheep-raising country, and sheep are capable of finding edible foliage where other, larger animals can't. So William went back into the sheep business, and it proved to be an excellent move on his part. It was the income from his sheep enterprise primarily that paid for his properties.

When William bought his first sheep and brought them to Menard, they were various colors. Mostly he bought Delaine bucks, which he purchased from an Ohio source. His plan was to cull out the speckled, spotted, and colored sheep until, over time, they were all white. White sheep made for a more profitable shearing business.

An Ohio breeder would send a boxcar of sheep to Menard from time to time, and the animals were auctioned off by the river. Once William gave two hundred dollars for the best ram in the carload. (According to Measuring-Worth.com, $200 would be worth about $5,200 today.) As a breeder himself who sold sheep to other ranchers, William strove to have the best strain of breeding animals available. To get the best breeding stock, you had to pay for the best, and William didn't mind paying the price. This animal was not only big but it also had great bone structure and produced a lot of wool. When they sheared the ram, one of William's sons weighed the wool on a fish scale. It was

Tools of the trade

Delaine sheep

in excess of thirty-five pounds, the scale's limit, and they still had more wool on the table.

Life was tough in west Texas in the early 1900s. On one occasion William and others, including his son Walter, went down to Del Rio, which was about 130 miles away, and bought a herd of mutton sheep from Pat Rose, a rancher with whom William had dealt before in another part of Texas. (Likely the same man he did business with in sheep when he first came to Texas.) This family had been known for generations as having fine animals. Once William completed his purchase, they drove the sheep on foot—130 miles—back to the ranch in Menard.

Since fences were scarce, ranchers could move their herds from one area to another most of the time, but there were problems on occasion. The sheep were led on foot, and a car was used to haul their supplies (like a chuck wagon). Other than the car driver, the rest of the group walked the whole way back.

On this particular occasion, not everybody along the way was hospitable. When the group arrived at a place just west of Menard, the owner wouldn't let them cross his land, so they had to drive the sheep on the road for a good distance. They finally herded all the sheep back home.

This kind of sheep movement, walking sheep for over a hundred miles, was a normal ranching practice.

## William's Love for Horses

William raised racehorses for a number of years when he first came to Menard. He always kept a lot of horses and sold many to the US Army for cavalry remounts, but he always kept about 150 mares on hand. Other than breeding for his first love—fast horses—he also bred for good saddle ponies. In addition, he liked to breed Percheron stallions with local Spanish mares to make a combination draft and/or stock horse. These were quite popular for farming operations and pulling wagons. Of course, he also bred mules too. William was a very competitive breeder, so if he intended to provide the market with the best, he had to be sure to breed the best.

    The horseless carriage was invented around 1900, but no one thought it would ever come to much. Those contraptions were far too loud, and in town they scared the horses in the street. They were loud enough with their primitive mufflers, but they were really loud when their engines backfired, which happened quite frequently. There were no service stations in the rural districts, and it seemed like the roads were too rough, rutted, and rocky for the skinny little tires, along with countless other mechanical problems. Throughout most of William's life, horses were king. Everything moved by horse and wagon, and it didn't look like that would ever change. After all, it had only been that way since the beginning of time.

    William's interest in fast horses never diminished. When the railroad first came to Menard in 1911, amid all the celebration, William still had an appetite for horseracing more than just about anything else. He had a freakish-looking horse that stood terribly poorly; he could stand naturally with all of his hooves in a number-two washtub. This animal certainly didn't look like much. Most pedigree racehorses stand with their legs

Mid-1800s spurs

William's son George at the headquarters ranch.

straight down at the four corners. But besides that, this horse's nose was so small he could eat out of a gallon bucket, so it didn't look like he could get enough oxygen to run fast in a race for any kind of distance. True enough, he might have been a poor-looking horse, but what he lacked in looks, he made up in speed.

William harnessed the horse with another one to his buggy and went to town, where he knew there were usually two other gentlemen who had stables of the finest and fastest horses around. One was a banker, whose bank was named after him, and the other one was a major stockholder in another local bank. William began chatting with them when he got to town, and at the right moment, he said, "I've got an old buggy horse hooked up to my hack out there that can beat anything you've got in your stables." They knew they had some fast horseflesh, and after taking a glance at William's pitiful-looking buggy horse, they jumped on the bet. William unhitched the horse from the buggy, replaced the harness with a bridle, and rode him bareback in the race. To the astonishment of the two men, William easily won the race and the purse. Although the outcome of the race made him some fast money that day, it didn't do much for William's banking relations, but he really didn't care.

Most Saturdays, William would go to Menard with one or two sons, if not the whole family, and he would usually tie one of his racehorses to the back of the buggy. By midafternoon, he would have a match race with someone on Main Street, smack in the middle of town. There were no speed limit signs for horses in those days.

Back then, the US Army would lease stallions to various ranchers and later buy the colts. The army had an interest in making good studs available to ranchers to breed so they could expect a steady supply of quality remounts in the future. One day William received the lease on a nice army stud.

Shortly thereafter, William and another man argued about the deal, and there were hard words and strong feelings behind them. Afterward, the man went around Menard with a big rock in his pocket, telling everyone he was going to use it to whip William. Of course, William heard about it all. So the next time he came to town, he saw the man in the street, went up to him, and asked him if he had anything on his mind. Irby McWilliams and Marion Whitley were nearby and saw William give the man a real whipping, barehanded, all by himself. After the altercation, McWilliams and Whitley asked the man why he didn't use the rock he had in his pocket. He replied, "I forgot all about it." However, that little bare-knuckles session seemed to settle the matter, because there were happily no further problems between William and the man.

William continued to breed horses and produce some pretty fast movers. From time to time he would load a few of his youngest and fastest horses on a train and take them up to meet his brothers in New York. All three of them loved racing horses. The Belmont Stakes had been run at Jerome Park in Westchester County, New York, since 1867, and it continued there until 1889. After Jerome Park closed in 1889, the Belmont was held at Morris Park in New York City until 1904. The Belmont was just once a year, but horseracing was quite popular and went on all the time. William and his brothers would race William's horses and split their winnings. Their correspondence included their ideas about which horses should be bred with which. Their racing colors were light green and white.

William's ranching and farming operations grew significantly, so he hired a number of ranch hands to meet the demand. Some lived in tents during harvest times. Although William had a number of men working for him over the years, certainly Blue Esteppe was a key man on the ranch. Esteppe was William's right-hand man well into William's retirement years, helping him with everything imaginable. He lived in a two-room house just southeast of the homeplace. It was originally built by George Menzies and his wife, Ella, and was where my dad, Perry Menzies, was

Pasture on the San Saba River

George and Ella's first home, ninety-four years later.

born. Blue was a major part of William's life and quite competent to help him manage the intricacies of a large, challenging, and constantly changing ranching and farming operation.

Calves at the feed trough

## Buying Land in Menard County

After buying his first two sections of land on September 3, 1887, for about $2.50 per acre, William worked constantly to improve the ranch. After moving the original home from the river to the top of the hill following the great flood of June 1899, William added a second story to the house. Many amenities were also added until it became a substantial country home. William and his brothers built a hydraulic press (or ram) to carry water from the San Saba River to the tank he erected on the hilltop. In addition to their large family home, he built four barns, a granary, several small outbuildings, and a shop with a forge.

With the homeplace, his ranching and farming operations in order, and his wife's industrious food distribution business going, William began buying as much land as his finances allowed. On January 4, 1888, he purchased 35 acres from his in-laws—the Jim Chastains—for $10 per acre. Jim Callan sold him 211 acres on the same day he bought the land from the Chastains. In 1904 he bought 30 acres from S. F. Bethell, and then he procured the Jackson property of 116 acres just east of the homeplace for $5 per acre. On November 25, 1904, he bought the place his brother Alex

George at the Gap ranch

The windmill and tank at the Gap ranch

had sold earlier to the Noguesses—1,200 acres—for about $7 an acre. Then in 1908 he added some school land, totaling about 320 acres at $6 per acre. Some thought he bid too much for the school land at the time, but he said, "It was good land and I didn't want to lose the bid."

With the river ranch well situated, on December 1, 1913, William bought the Gap ranch, consisting of 3,200 acres in the northwest end of the county on Wilhelm Lane from Louis and Mary Ellis for $7.50 an acre. A home was soon built on the property, and the first family William had living there and running it was his eldest son, George, and his wife, Ella. Then came World War I and the Great Depression. Times were really rough, but in 1924, William bought another 2,000 acres out at the gap from Bob Sears at $12.50 an acre plus another 160 acres just west of it. Along with it, as part of the deal, he also bought six hundred head of sheep at $12 a head. In all, William and Letha Ann purchased over 8,500 acres of land in their lifetime.

## Financial Arrangements

To finance these purchases, William borrowed a good deal of money from his brother Alex in New York on an interest-only basis. Their agreement was

George Menzies operates a spreader at the Gap ranch

that he would pay Alex 10 percent interest until either he or Alex died, and then the principal would be forgiven by virtue of all the interest payments he was to have made over the years. This was a popular idea at the time. The lender would actually get all of his money back in ten years, and the borrower still owed the principal amount until one of them died. Money from this arrangement was a large part of the seed money William used, along with his own capital, to buy some of his land initially. He also borrowed money from his brother George to buy another large tract. William's brother Alex was a bachelor, but in his latter years he had a mistress named Mary. She was originally a nurse or an assistant to him, and he subsequently married her after his mother, Agnes Craigmile Menzies, died. For some reason the family was not told about the marriage until several years later.

## Remaining Flexible in the Ranching Business

As is the case in just about any business, William had to be flexible, alert, and willing to change his operation instantly due to constant changes in the market, weather, and other conditions. For a long time, William also raised mules and sold a good many of them to the army. Mules really pulled farm implements and wagons far better than horses.

There's an interesting story about William's youngest son, Walter, when he was about five years old. William was having his men round up the mules in the back pasture. Every one of the mules happened to be perfectly in tune with their species' reputation for stubbornness. The hired hands were using Walter as a messenger to take information to William at the house. Little Walter wasn't supposed to be working, though. The hands were lagging back to see to it that the herd of mules didn't turn around and scatter, but the mules headed for the barn, leaving the men behind. Walter kept his horse right on the mules, running the herd into the corral, and then he shut the gate behind them. Once he had rounded up the whole herd, he jumped down, and locked the gate behind those agitated mules—all by himself. That was a courageous feat for a five-year-old.

With the advent of gasoline-powered vehicles, the value of mules dropped off quite suddenly. At that point William sent word to all the high school boys in town that he would sell them his mules for five dollars apiece if the boys would ride them off the place. He received a lot of five-dollar bills initially but ended up giving most of them back because hardly any of the boys could ride them. After a fashion, the boys would return to the ranch house, walking back up the road with their heads

Anne Crawford (later Menzies) with some kid goats

Shearing sheep in the 1930s

hung low, dusty, dirty, and well worn from their attempt to ride home on a mule at a bargain price. William didn't hesitate to give their money back.

He ranched in Menard, raising registered Hereford cattle, horses, sheep, mules, and goats and always maintaining a keen interest in racehorses. He was credited with bringing the first jockey uniform into Menard County. His large animal production created a passion in him to do a lot of diversified farming as well. He needed to grow feed for his livestock and tried to produce whatever cash crops he could beyond that. During his more than fifty years of ranching in Menard, he was considered a Texas leader in progressive agriculture. Among other things, he pioneered new techniques in cattle, sheep, and goat production.

## Entering the Goat Business

In 1912, William entered the goat business by buying a bunch of muttons from Dick Godfrey. He also bought twelve hundred head of goats from what was known as the Lawrence Ruff place on the river. Max herded that bunch of goats back to the headquarters ranch because they still didn't have any fences in that area. Right after that William bought a bunch of mixed goats that were being pastured out at the Gap ranch. He built up his herd by buying heavy shearing billies. This proved to be a good move, because goats are still being run productively to this day by his descendants all over the county.

On the 3,200-acre headquarters ranch on the east side of Menard County, William usually ran about 800 high-grade Delaine sheep with registered Delaine bucks. He also had about 500 head of high-grade Angora goats with registered billies, along with about 100 head of high-grade Hereford cows and registered bulls. On his 5,000-acre Gap ranch on the northwest side of Menard County, he ran 1,500 Delaine sheep along with registered bucks. He also had another 700 Angora goats with their billies, and another 150 high-grade Hereford cattle with registered Hereford bulls. Three miles of the San Saba River ran through the headquarters ranch, furnishing an abundant supply of water for farming and ranching operations. William had 180 acres of land in cultivation there most of the time, raising various feedstuff, such as higuerra and milo maize. He was constantly improving both ranches with homes, silos, wolf-proof fences, corrals, wells, windmills, water tanks, barns, sheds, and feeders.

Angora goats

*Ranching in West Texas*

William's son Alex in the cow pen

Inside the feed barn

William kept up with all the latest ranching practices and was a member of the Sheep and Goat Raisers Association, the American Hereford Cattle Breeders' Association, and the Texas and Southwestern Cattle Raisers' Association.

## Chapter 7

# Pioneer Days

To UNDERSTAND THE LIFE and times of William Menzies, one needs to understand the state of Menard County and what was going on there just prior to and immediately following his arrival in 1877. In a nutshell, this was a time of great struggles and much bloodshed encompassing several wars between nations and constant clashes with Indians, robbers, bushwhackers, and wolves. Many lives were lost and a large amount of blood was spilled in the San Saba River Valley. This whole area was a rough and dangerous place. Once you know what the region was like, it is no wonder why William never went anywhere without his two pistols. Regardless of the progress the Menzies family was making as pioneers who were developing their ranches and raising their children, Menard County was at the outer edge of civilization. It

was not a place for the fainthearted. In fact, the most notorious area of the region, and certainly in Menard County, was a place called Robbers' Roost. And this den of thieves bordered on the southeast corner of William's headquarters ranch. In order to survive in the bush on your own, eight miles from town, you had to be alert, extremely cautious, and at the ready *all the time.*

During most of William's life there was no large sheriff's department in Menard County. The time was so rough with roadside bandits, horse rustling, cattle- and sheep-stealing feuds, and the like that William never went to bed at night without a pistol under his pillow. And this remained so almost until the day he died. It was well known in Menard County that William sold horses and other livestock in large numbers, grains by the wagonloads, and Letha Ann sold a large volume of produce, so most folk knew that William and Letha Ann had to have cash on hand at the headquarters ranch just to keep the operation going.

For this and other reasons, William and Letha Ann had several rifles and shotguns in the house as well as pistols. William's favorite two pistols were both made by Smith & Wesson. One was a pearl-handled, nickel-plated, engraved, Russian caliber .44 six-shooter—one of the nicest guns you could buy back then. He also had a matching but smaller nickel-

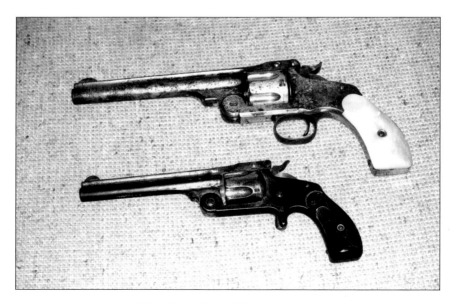

William's two Smith & Wesson revolvers

plated .38 caliber, five-shot revolver called a Bulldog. It was small enough to fit in the top of his trousers and under his belt. Since he always wore a vest under his coat, which covered the handle, most people were never aware that he carried it so often. This pistol was peculiar because it was manufactured without a trigger guard, thus it would slip in and out of a belt without any problem and you could get to the trigger quicker in an emergency. On his wagon trips, he always carried the .38 in his belt and kept the .44 under the wagon seat. If five shots from his .38 were not enough, six from the .44 usually could be relied upon to finish the job.

To better understand this territory, what follows is a brief account of went on in the San Saba River Valley just prior to and including the time in which William and Letha Ann were ranching there.

## Twenty-Six Indian Tribes

For several hundred years—according to John Kniffen, the resident Menard expert on Indian lore and a retired professor from Texas Tech, who has studied Indians his whole life and was an abundant source of knowledge to me—more than twenty different tribes of Indians made annual pilgrimages to the San Saba River Valley in the fall, usually during the

An Indian scout

month of October, to harvest native green pecans. While in the valley they ate only wild game and mussels from the river. They carried the pecans north to munch on while buffalo hunting during the winter. The pecans offered a naturally preserved source of protein and fiber. To eat them, the tribes did not crack open and shell the pecans. Instead, they crushed them whole and ate shells and all. Amazingly, though there were

A sample of a native pecan (left) and today's improved variety (right)

bitter rivalries between the tribes, they rarely fought with each other while collecting pecans in the valley. (White pioneers, however, did not get off so easily; the tribes always attacked the Anglos.) Pecans were so bountiful, there was apparently no call for the tribes to fight over them. Most of the time, the tribes hunted buffalo farther north, and this animal provided Comanches and Apaches with nearly everything they needed. The meat was their main food source, but they were able to make tools from the bones and bindings with the tendons. Hides provided clothing and lodge coverings. Most Indians preferred deer hide for clothing, because it had a softer

Pecan trees line the San Saba River around Menard

# Pioneer Days

A closer view of some pecan trees along the river

texture. They also gathered food such as clams from rivers and berries and fruit from the shrubs and trees.

The Spanish considered Apaches to be the most warlike of all the tribes. This remained true until Comanches emigrated south from the Colorado area and the central United States. When Comanches adopted the Spanish horse culture and black-powder rifles, they became a force to be reckoned with as the "lords of the plains."

Without exception, every tribe was impacted negatively by interaction with the Anglos. Their buffalo food source was decimated, their culture was destroyed, and their people were polluted by alcohol as well as highly contagious and often fatal diseases. But in the mid-nineteenth century, white Americans were infused with the spirit known as Manifest Destiny—a belief that the whole land mass of America was to be seized and developed out of a kind of providential calling.

## The First Spanish Settlements

In 1757 the Spanish had dreams of a new Spanish Empire in Texas and beyond, starting with the San Saba River Valley. The Mission Santa Cruz de San Saba was founded by Father Alonso Giraldo de Terreros specifically

Entrance to the Presidio de San Luis de las Amarillas

to convert Eastern Apaches to Catholicism and ultimately make them Spanish citizens. The mission concept was instituted with the hope of peacefully absorbing the Indians into the Spanish culture and converting them into a labor force. There is no doubt that many of the Spanish and their monks had a true zeal for evangelism, but converting Indians into laborers went against basic Indian culture. Converting them to Catholicism was, likewise, too great a leap of faith for most tribes and could not be done in a single generation.

Concurrently, Col. Diego Ortiz Parilla built the Presidio de San Luis de las Amarillas to protect the mission and to secure the area. In addition to protecting their fledgling interests there as well as Spain's general presence throughout the Southwest, the fort was also a deterrent to any claims by the French to this vast territory. After all, the French had already gained a dominant presence in New Orleans and the Louisiana area.

Initially, both the fort and the mission compounds were constructed of timber and their stockades were enclosed with logs. Later these were replaced with stone, affording both sites more strength and permanence. Experiences over time with Spanish soldiers had taught the priests and monks to build their missions about three miles downstream and on the other side of the river. This was done to keep the soldiers from having

*Pioneer Days*

Walls of the presidio

too much influence over the Indian converts or from seducing the Indian women.

Although Apaches frequented the San Saba mission, they never actually entered into residency there but always promised to return later to see the monks. Meanwhile, the Spaniards built an irrigation ditch system, which was later improved, making Menardville into a fertile farming valley. But befriending the Apaches, as the monks did, gained the Spaniards something they didn't expect: the wrath of the Comanches.

On March 16, 1758, a confederacy of several northern tribes, mainly Comanches and Wichitas, some two thousand strong, descended on the mission and killed two priests and six other Spaniards, and wounded several other men. Most of the soldiers at the presidio were either away or on other assignments. The small detachment finally sent from the fort to help the mission was quickly driven back by Indians. This was the first hostile conflict between Europeans and the Comanches in Texas, as well as the first time Spaniards had engaged a large number of Indians who were armed with weapons obtained from the French. The Europeans had brought guns and horses to the Indians, and now they had to meet the Comanches in battle armed as equals. Over the ensuing years, Comanches regularly attacked the presidio and the settlers in the area. The Spanish mounted numerous retaliatory attacks, and thus both sides suffered heavy losses. The Comanches continued to harass the settlers in this area for another hundred years. Then, years later, both the mission and the presidio were abandoned. The monks were frustrated that all their efforts over all the years had resulted in only one conversion and baptism.

Though the Spanish had poor results with the native tribes, they continued to strive for dominion over all of Texas through other missions and presidios. They never dreamed that Anglos would start streaming into the territory from the east as early as 1815, but they did. Initially, they reluctantly permitted the Anglos' immigration as part of a desperate attempt to develop Texas and make it pay for them economically.

## Mexican Independence

In 1812, three hundred years of Spanish control and colonialism ended in Central America. Mexico declared its independence from Spain and went into a three-year period of uncertainty as a republic and then, under the leadership of Augustin de Iturbide, as an empire. In 1824, Mexico's leadership set up a federative republic very similar to that under the Articles of Confederation that the United States used to govern and bridge the gap between English colonialism and the adoption of the US Constitution. During this time, they allotted the most important powers to individual states. As Anglos continued to flow into Texas from the eastern United States, Mexico became increasingly suspicious of a US plot to annex Texas. There was no such plot at that time, but trouble was brewing in the huge Texas caldron where nationalities, languages, and cultures clashed daily.

The Indians were a constant problem to the pioneers in the San Saba River Valley, terrorizing settlers, raping women, burning settlements, stealing livestock, and killing isolated Anglo ranchers. They started wildfires to keep the trees from growing too much in that part of the country, probably to make life hard on the settlers.

Jim Bowie, an Anglo immigrant, married the daughter of the Spanish vice governor and received a huge land grant as well. This property consisted of fifteen eleven-league grants of Texas territory. (A league was four thousand acres.) However, the land grant wasn't worth much if the Indians killed your family, stole your animals, or burned you out of your home and ranch. Thus Bowie befriended the nearby Apaches and also lived with them for a time. But by befriending Apaches, he brought on the wrath of Comanches.

According to legend, just before going to San Antonio for the last time, Bowie stole a large amount of silver from some Apaches and hid the major portion of it in an underground cave. Menard County has more underground caves than any other county in Texas, and Bowie was always warring with the Indians on the outskirts of Anglo civilization. Just after hiding the silver, he had his last battle with the Comanches about a mile northwest of what later became William Menzies's ranch. Bowie survived that fight and headed for San Antonio, leaving the silver behind. Though it has been searched for diligently on numerable occasions and sometimes quite elaborately, the lost silver and lost Indian silver mines have never been found.

Bowie knife

The battle of the Alamo, where Bowie gave his life for Texas freedom, took place in 1836. The site of his last battle with the Comanches was at what is now called the Five Mile Crossing (five miles east of Menard on the San Saba River). There is a large amount of historical data to support this as the site. Also, the latest archeological digs and finds—as well as the large number of artifacts and relics found there by Texas Tech University researchers—mark this as the place.

During the early and mid-1800s, it was standard Indian practice to kill and scalp all Anglo adults but to adopt young children and indoctrinate them into their culture. When the boys were old enough, they received instruction and were initiated as braves. Anglo boys who later returned to white civilization rarely fit in and wanted to return to what was left of their tribe. They had grown accustomed to a much easier way of life, where women did most of the work and men simply killed wild game and raided Anglo settlements. The men actually did very little in Indian culture. Several other problems kept them from integrating back into Anglo society. They lacked even a general education, had no language skills, had learned no trade, and had no idea how to farm or husband animals other than horses. Of the many German children who were captured by the Indians and indoctrinated in this manner, the last one was released in 1877.

## The Texas Revolution

Another important chapter of the San Saba River Valley was planted deep in Texas history during the years while the whole state was in Mexican hands. When the Mexicans took control of Texas from Spain in 1821, they quickly became nervous about the steady influx of English-speaking Anglos. A further concern was their having ties and loyalties primarily to the United States. At the time, most of Mexico's people were content to occupy the more hospitable land south of the Rio Grande. Few indeed wanted to face the constant Indian and outlaw attacks that were occurring in the Texas territory. The Mexicans had a viable option while the huge numbers of Anglos flowing in from the east were desperate for any piece of real estate. Very quickly, the Mexicans in Texas were outnumbered by the Anglos ten to one.

On April 6, 1830, Mexico's Congress passed a law banning further Anglo immigration while encouraging the same from all other nations. This created a rift with the Anglos, because it meant their friends and family members back home could not join them. The new law allowed convicted Mexican criminals to homestead in Texas as part of the effort to dilute the Anglo population. Anglos also were denied the right of petition or assembly. The Mexicans tried to discipline and control the American colonists in several ways. They abolished slavery, established military forts, levied taxes, and finally declared martial law in a desperate effort to disarm the Texans. They discovered that trying to take guns away from Texans was a terrible idea.

Inevitably, fighting erupted in many places, but most notably at Gonzales on October 2, 1835. The Mexicans were better armed, better equipped, better fortified, better financed, and were able to fend off the first attacks. On November 3, 1835, several Texans met at San Felipe and formed a provisional government. Unfortunately they couldn't agree on very much, including a decision to declare independence from Mexico. In any nation contemplating independence, there are always the "haves" with land, con-

The Alamo

nections, and contracts with the current government who want to stay and the "have-nots" who believe independence is their only option. However, all this indecision soon brought tragic consequences. Although the Texans captured San Antonio and forced the Mexicans to withdraw to Mexico, this was to be the Texans' last victory for several months. On March 2, 1836, Mexican dictator Antonio Lopez de Santa Anna began amassing his army to lay siege to the Alamo, a time-worn mission that had been made into a makeshift fort. Concurrently, however, other Texans were meeting at Washington-on-the-Brazos to declare their independence from Mexico and establish the Republic of Texas.

The fledgling nation of Texas could not have chosen a more tragic, bloody, or heroic beginning than the battle of the Alamo. The compound was defended against all odds initially by only 150 men. On March 1, another 32 Texan reinforcements slipped through Santa Anna's lines and into the Alamo. Those in leadership that day were most notably Col. William Travis, the commander, along with James Bowie and David Crockett. On March 4, defying Santa Anna's demands for unconditional surrender, about 4,000 Mexican troops launched an assault against the Alamo. The Texans fought valiantly with rifles first, bayonets next, and finally with knives and whatever they could find in hand-to-hand combat to the death. The battle cost the Texans about 187 men, but it cost the Mexicans an estimated 1,000 to 1,600 lives. What the Texans lost in blood that day, the fledgling republic gained in a multiplied fashion in resolve, emboldening and focusing their will as a nation to win their precious independence.

Right after the Alamo fell, there would still be yet another huge price to pay for independence. As it happened, Col. James Fannin commanded about 300 Texans at Goliad and was terribly outnumbered by a Mexican force. In desperation, Fannin finally ordered a retreat. However, he was not actually able to retreat until March 19, and his force was soon overtaken by the Mexicans that afternoon. Hopelessly outnumbered in the tragic battle that ensued, Fannin finally surrendered. He and his troops were taken back to Goliad, and Santa Anna ordered their wholesale execution on March 27, 1836. History clearly shows that the blood of these two defeats produced the fruit of an overcoming and victorious political will among the Texans. The heroic defeat at the Alamo and the cold-blooded massacre at Goliad galvanized the Texans as a nation and gave them determination and a new battle cry: "Remember the Alamo! Remember Goliad!" They soon used it triumphantly when the two forces met in battle for the last time at San Jacinto.

The battle of San Jacinto was fought on April 21, 1836, by the San Jacinto River, near what is today Houston, Texas. The Texans under Gen. Sam Houston had been in retreat for several days. But on that fateful day the Texans were able to defeat the much larger Mexican army with a bold surprise attack initiated while the Mexicans had let down their guard for a siesta. In the twenty-minute battle that ensued, 630 Mexicans were killed and 200 were wounded. In addition, Santa Anna and 700 men were captured. Only 9 Texans were killed and 30 were wounded, including Sam Houston. He took a bullet in the leg and lost a lot of blood. But on that day, Texans won their freedom, and Texas became a sovereign nation.

Although the story of the role played by the fledgling four-ship Texas navy in winning freedom for the republic is little known, it should be revered and never forgotten. Most historians now agree that the victory at San Jacinto was made possible by the strength, activity, and threat of the Texas navy. The small fleet was headquartered at Galveston and charged with the mission of protecting the entire coast from a Mexican seaborne invasion. If the Mexicans had landed additional troops and reinforced Santa Anna's army, such a force likely would have overwhelmed, trapped, and annihilated Houston's ill-equipped but valiant minuteman army. But the Texas navy prevented that from happening. And after the battle of San Jacinto, Santa Anna was imprisoned on one of the ships of this small Texas armada. However, the Texas navy was too hastily decommissioned shortly after Santa Anna's surrender, because the fight for Texas independence was not yet over.

Even after the Texas victory at San Jacinto, Mexico stubbornly refused to acknowledge Texas as a nation. To make matters worse, most of the citizen-soldiers of Houston's army returned to their homes to take care of their families, farms, and ranches. Although Texans had declared their independence, their republic was not yet organized to the extent that it had a properly functioning government, much less a tax base to support a standing army. Thus the fledgling republic faced renewed threats of even greater land and sea invasions by huge, overwhelming, and continually mounting forces from Mexico. Though soundly defeated at San Jacinto, Mexico had not taken its eye off of the prize of the land and natural resources of Texas. To meet this growing threat, the republic commissioned a second Texas navy. The eight ships of this new navy were commanded by Cdr. Edwin Moore. Over the next three years his fleet harassed and outmaneuvered the Mexican navy from the muddy waters of the Rio Grande to the Yucatan Peninsula. This campaign culminated in a final victory over the Mexican navy at the battle of Campeche. Only then did Mexico abandon its plans for invasion—even temporarily. Mexico did not, in fact, give up its efforts to reclaim Texas until many years after the republic

The ensign of the Texas navy

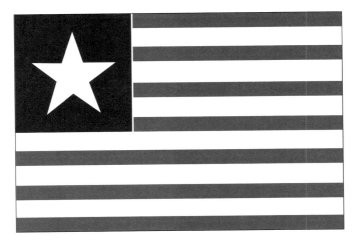

had been annexed by the United States. The Texas navy, however, continued its mission of providing protection for the republic's thousand miles of coastline for another ten years.

During his presidency in Texas, Mirabeau B. Lamar ordered Moore's fleet to assist the Yucatan rebels in their bid to challenge the Mexican government. When Sam Houston was reelected president of the republic, he reversed this policy because he believed it to be an expense the fledgling nation could ill afford. He also thought it was an unnecessary and unwise provocation of Mexico's leaders. Houston quickly ordered Moore's navy back to Texas, but Moore refused. In reply, Houston deemed Moore to be a pirate and encouraged all other ships at sea to attack and destroy Moore's flotilla. This brought Moore to his senses, and the Texas navy soon returned to its home port.

To his credit, Moore had a fighting spirit, which, believe it or not, is many times not easy to find among soldiers and sailors. For those unfamiliar with military life, it is important to note that a commander charged with taking a hill, a city, or a country must then make a commitment in his heart to accomplish that task at all costs, at the risk of his own life and limbs. The commander as well as his troops become fully committed to the assigned mission and will then achieve their goal, no matter what. Once that commitment has been made, it is difficult for them to back off even for a minute, let alone do a 180-degree turn. The classic example of this dedication is President Harry Truman's troubles in getting Gen. Douglas MacArthur to shift from offensive actions to a restrained defensive position during the Korean War. Truly, MacArthur should have followed the orders of his commander-in-chief. In the end, Truman had to fire him, but that doesn't mean MacArthur wasn't a great general.

The difficulty in finding fighting generals plagued President Abraham Lincoln as well. He went through ten commanders during the Civil War until he found the fighting general he needed—Ulysses S Grant. Anyone can wear a uniform and even graduate from a military school, but to win a war, commanders at some point and on many occasions must close with the enemy. They must be willing to lead their men in a combat assault directly into the full physical strength and firepower of the enemy. Not many commanders have the stomach for that.

Just because Houston's army had defeated Santa Anna at San Jacinto, and he promised them that their independence did not mean Mexico was willing to just walk away. They wanted Texas back, and they wanted it badly. Without the vital role played by the Texas navy, however, there would almost certainly have been another Mexican invasion of Texas. Instead, for the next decade, Texas remained an independent republic. During all that time, Texas did everything it could to be annexed by the United States, but a number of US politicians believed that adding another slave state to the Union would have upset the balance of power. The spirit of Manifest Destiny, however, overcame those reservations, and on December 29, 1845, the US Congress voted to annex the Republic of Texas. The small but powerful and critically important Texas navy was finally absorbed into the burgeoning naval armada of the United States.

## US Dominion of the San Saba River Valley

All during the nineteenth century, US settlers arrived in Menard and used the old presidio compound as a fort to stave off Indian raids. In response to these constant Indian attacks and to promote the westward expansion of the country by Anglo settlers, the US government erected more than twenty forts across Texas. One of these was Camp San Saba, which is seventeen miles west of Menard, on a commanding knoll above the river and just inside the county line. Later the stronghold was renamed Fort McKavett, after Capt. Henry McKavett, a casualty of the Mexican War. The post was and still is quite picturesque. Gen. William Tecumseh Sherman called it "the prettiest post in Texas."

On March 4, 1852, Col. Thomas Staniford arrived at Fort McKavett with an infantry regiment. He was charged with strengthening the fort and scouting out and eliminating the Indian tribes still in the area. In 1859 the army abandoned the fort because many believed they had the Indian situation under control.

However, in the 1860s, during the Civil War, Fort McKavett was reactivated because of renewed Indian attacks, including one that caused the death of William McDougal. At the end of the war, the fort's secondary mission was to keep an eye on abolitionists and secessionists who some people thought might launch another insurrection. The county was still sparsely populated, but this was a growing land. After all the destruction in the South during the war, a new wave of immigration was about to hit the state of Texas, including Menard County. From 1860 to 1870 the state grew in population from 37,363 to 61,125.

Many well-known soldiers were stationed at Fort McKavett over the years. Among them was Gen. Ranald S. "Bad Hand" Mackenzie, who became a renowned Indian fighter. Mackenzie was ranked first in the West Point class of 1862, and he fought for the Union throughout the Civil War at the battles of Second Bull Run, Antietam, Gettysburg, the Wilderness, Spotsylvania, Cold Harbor, Trevilian Station, Fort Stevens, Cedar Creek, Petersburg, Five Forks, and Appomattox Court House. He began the war as a second lieutenant and received seven brevet promotions, ending the conflict as a brevet major general of volunteers when the smoke cleared from the last battlefield. Mackenzie was a harsh disciplinarian and not well liked by his soldiers, although he was highly regarded by his superiors. He was wounded six times in combat, notably at Second Bull Run, Gettysburg, Opequon, Cedar Creek, and during the fighting at the Jerusalem Plank Road during the Overland campaign. The last wound cost him two fingers and was probably the source of his nickname, "Bad Hand." After the war, Mackenzie remained in the regular army and his rank reverted to captain in the corps of engineers. In 1867 he was promoted to colonel and dispatched to the Texas frontier, where he remained for the rest of his career, commanding the Forty-First Infantry, a Buffalo soldier (African American) regiment. While some officers were not eager to command black troops, Mackenzie had no reservations regarding the men under his command. There was no shortage of African American recruits. After the war, jobs were hard to

Fort McKavett headquarters building in 2010

*Pioneer Days*

come by. Thirteen dollars a month and free room and board was a pretty good deal.

In 1871 Mackenzie was named commander of the Fourth Cavalry Regiment and led his men at the battle of the North Fork, which is in the Llano Estacado of west Texas. Later that year Colonel Mackenzie was wounded again, this time by an arrow in his leg. In 1874 his command participated in the Red River War and defeated the combined warring tribes at the battle of Palo Duro Canyon, which was well north of his headquarters at Fort Concho in San Angelo. In 1876, after the massacre of George Armstrong Custer's Seventh Cavalry, Mackenzie's command defeated the Cheyenne at what became known as the Dull Knife Fight, which helped to end the Black Hills War. Afterward he was appointed the commander of the District of New Mexico in 1881. He was promoted to brigadier general in 1882, and assigned to the Department of Texas in 1883. During his career, he was promoted seven times for bravery.

Mackenzie decided to plant his roots and live out his last days in Texas. He bought a ranch and became engaged to be married, but his behavior suddenly became erratic, possibly due to a head injury he sustained by falling from a wagon while at Fort Sill, Oklahoma. Due to this instability, on March 24, 1884, Mackenzie retired from the army and subsequently died at his sister's home in New Brighton, New York.

Among the other notable soldiers to be stationed at Fort McKavett were Col. Abner Doubleday—the legendary creator of the game of baseball (though he never claimed the honor), the Union officer who fired the first shots from Fort Sumter during the first battle of the Civil War, and the temporary commander of the Union army at Gettysburg until George Gordon Meade arrived on the scene—and Col. Adolphus Greely, a polar explorer and a cofounder of the National Geographic Society.

Fort McKavett was closed and recommissioned again and again due to recurrent Indian raids and attempts by various outlaw bands to get the upper hand in the territory by attacking the settlers. Finally in 1883 Fort McKavett had fulfilled its purpose and was officially decommissioned, abandoned by the government, and taken over by settlers.

## First Anglo Woman Born in Menard

In 1852 Clara Shellenberger was born in Menard County. She was the first white person to be born there. Previously only Indians and Spanish settlers had populated the land. Quite coincidentally, Shellenberger was also involved in the last big Indian raid in the county as well.

It the early morning of August 6, 1866, William McDougal and his wife, Elizabeth; her fourteen-year-old daughter, Clara (Elizabeth had been previously wed); and a son-in-law were living at the McDougal homeplace about two miles east of Fort McKavett. William and his son-in-law left home and crossed the San Saba River to search for a stray cow. Elizabeth and Clara remained at home and attended to their daily chores, taking advantage of a beautiful, bright summer day. Elizabeth was airing out the household bedding. She had brought feather beds with her from Pennsylvania, and she had several blankets, quilts, and pillows as well. There were places on their snake-yard fence to air things like this. Later that morning, she sent Clara to the vegetable garden to pick some roasting ears for dinner (they called the noon meal *dinner* back then).

Clara was a sturdy young lady. As she headed back to the house with the corn, she noticed a group of Indians on horseback crossing the river. Clara was nearsighted, so she could not see the Indians until they were quite close to her. When she realized what was happening, she dropped the corn and ran toward the house. Just as she tried to open the gate to the yard, a warrior rode her down and pierced her upper shoulder with his spear. When Elizabeth heard the commotion, she grabbed a gun and ran into the yard to rescue her daughter. She dragged Clara frantically into the house. After she removed the lance, she laid Clara on her back on the floor. Elizabeth then put on a man's hat, grabbed a rifle, and started talking in a loud, gruff voice, trying to give the impression of a man's being there. She started firing her rifle at the Indians from various positions around the cabin, because she wanted the war party to think there were more than just two women in the house. While it is doubtful that her deceptions had any effect on the war party, they never did try to come in the cabin. The warriors only circled the house, yelling and shooting their rifles. Some of them ripped and emptied all the feathers from the beds and pillows. Both Kiowas and Comanches made up the war party that day.

As Elizabeth was tending Clara's wounds, she looked out the window and saw her husband's riderless horse returning to the corral. She feared the worst. Later she found out that William had seen the war party and tried to hurry home to the family. But while he climbed the bank to get to the house, he was shot in the back with an arrow and killed.

Pioneer Days 97

About two hundred Indians comprised the war party that day. They gathered whatever loot they could, as well as all the horses and cattle in the vicinity—except for one horse that was saved when its owners brought it inside their house during the raid. It took several days for some of the horseless settlers to get to any other settlements and borrow some horses. By that time, the Indians had divided the animals among themselves and had departed the area in three separate groups. The settlers determined it was not feasible to pursue the Indians, considering their numbers and the condition of the families who had been attacked.

For about a year Clara was an invalid because of the spear wound, and it never really healed properly because her ribs had been severed from her backbone. She did, however, survive. Later she married a man named Mergenthaler, who himself had been wounded three times during the Civil War. In 1866 he reenlisted and became the mess sergeant at Fort McKavett. There he met Clara and married her. Upon his discharge from the army the second time, he suddenly and mysteriously left for New York City. Clara remained in Menard and worked as a laundress. When she had saved enough money, she went to New York, found her husband, and brought him back to Texas. She was driven by that same indomitable Texas spirit that inspired so many of the pioneers. The Mergenthalers settled in San Antonio where they built a chain of ten cigar factories, ranched, and raised a family. Clara enjoyed a full life and died in January 1931; her husband died the next year.

Texas heat

## Andes Murchison, Trail Boss and Rancher

Winning the West took fearless men who were willing to commit to long trail drives in the face of great danger to get their cattle to market. Some had to go from Del Rio as far as four hundred miles to the rail head at the Red River border between Texas and the Oklahoma Territory. Cowboys braved scorching summer heat and terribly cold winters while choking on dust and fighting fatigue, as well as constantly risking attacks by Indians and outlaws. Quite often trail bosses had to give a few steers to an Indian chief to buy safe passage. Although most trail drives were in warm to hot weather, winter runs were occasionally warranted. These wintertime drives tempted many who hoped to be the first herd to get to the railhead and catch the spring market. On one such winter cattle drive, a blue norther froze half a herd to death overnight.

Pioneering this territory called for young men like A. H. "Andes" Murchison. He was born on March 5, 1855, in New Braunfels, Texas, the same year William Menzies was born across the Atlantic. Murchison was one of eight children born to Capt. Dan and Mina Wilhemina Murchison. Dan came to Texas with a company of soldiers from Tennessee and fought at the battle of San Jacinto under Sam Houston. When the war with Mexico was won, Murchison brought his family to New Braunfels, where he worked as a surveyor. He also surveyed the city of Fredericksburg and later came by wagon train to the San Saba Mission to stake out a claim to various tracts of land he purchased directly from the state.

Once young Andes was old enough, he went to work on the cattle drives. As an adult he stood over six feet tall, and many said he was a giant of a man. He first hired on as a drover and then as a trail boss, making twelve trips up the Chisholm Trail. Other cattle drives took him to Dodge City and Abilene, Kansas, and Clayton, New Mexico.

Andes hired a lot of hands over the years for these drives. One time he hired a quiet young man from a well-to-do Northern family. At one point on the drive, this young man was helping Andes calm the cattle during a thunderstorm, and he was struck by lightning, which instantly killed both him and his horse. They had to bury him on the trail. A year or so later his family came to Texas, exhumed the body, and carried the remains back north for a proper burial.

After Andes's 1885 trail drive, having no idea that it had been his last, he wintered in Menard. He thought he had earned enough from all the cattle drives he had led to go into cattle ranching with his brother Ed, who controlled a lot of land left to them by their father. Their interests grew into a huge operation. It encompassed two large ranches—the Rattle Snake Ranch and the Forked Lightning Ranch—and a general store in Menardville. Their holdings grew to include the Paint Creek Ranch in Edwards County and another ranch in Kimble County, at a little town called Telegraph. They bought the ranch in Telegraph in order to harvest the cedarwood for fence posts to use at their other ranches. Andes was known as "Uncle Andes" to most folk, and he remained a businessman operating his mercantile business in town and a rancher until his final days. He had a number of friends, and very late in life he even married. When he got older, he hired some ranch managers to take care of his livestock.

Andes always rose early every morning and dressed in knee-high boots—he didn't like short-top boots people had begun to fancy. Usually he wore a three-piece suit, a string tie, a long coat, and a wide-brimmed hat. In his old age, he just supervised his ranch managers and the mercantile store manager in town. He always stood straight as a ramrod until his very last days.

Andes and William Menzies became fast friends who shared a first-generation Scotch ancestry, a self-reliant spirit, and a common vocation as ranchers. Both men seemed to "bore with a pretty big auger," and this gave them much to talk about when they met along the path of life in Menard. Andes lived to be ninety-seven and truly left his mark on Texas.

## First Anglo Male Born in Menard County

In the summer of 1859 Lucy P. D. Robinson and her family arrived at Fort McKavett by covered wagon. She was accompanied by her fifteen-year-old daughter, Lucy Ann, and her sons, Joshua and Robert Robinson, and their families. Formerly from Boston, Massachusetts, the Robinson family came to Fort McKavett as part of an ox-drawn wagon train from New Braunfels. Capt. Dan Murchison's wagon train preceded them there by a day and a half's travel time. Both wagon trains had the same destination: the San Saba mission. Both families had decided to purchase land there, and both had been issued patents to buy the land from the state of Texas at twenty-five cents per acre. Mrs. Robinson bought more than two thousand acres in and around Fort McKavett, on the Concho River at the mouth of Kickapoo Creek, and in both McCullough and San Saba counties. Murchison bought land in and around the old mission, and that land would be the first portion of the massive ranching operation that would subsequently be carried on by his sons, A. H. and Dan Murchison.

Lucy Robinson and her children and their families lived in the old headquarters building at Fort McKavett until she died in 1866. A good part of the family was still living there when the fort was reactivated and reoccupied by the US Army in 1867.

Lucy also bought large herds of cattle to graze on their vast domain. The land was all open range at that time; none of west Texas had been fenced back then. She hired Jack Livingston to oversee her cattle operation, and under his leadership, the herds increased significantly. In spite of frequent Indian raids, the family flourished in the San Saba Valley.

In 1862, George Roberts, an adventuresome Englishman turned cowboy, was just drifting through and happened to meet Lucy Ann at Fort McKavett. They fell in love and were married. On July 5, 1863, George William "Willie" Roberts Jr. became the first white male to be born in Menard County. (Willie was also my cousin Scotty Menzies's grandmother's stepbrother.) George Sr. and Lucy Ann also had a daughter, Florence Elizabeth.

All was well with Willie, but when he was two years old he developed a very high fever, which they called "white swelling," or infantile paralysis. Today, we call it polio. Both of his legs became paralyzed. His uncle, Dr. Bob Robinson, treated him the best he could, and they say that he probably saved the boy's life. Still, Willie's left foot and leg were twisted and shriveled until they were never much more than a hindrance. His right leg was also affected, but not as badly. Willie hopped around with a crutch under his left arm. Doc Robinson later sent him to a doctor in San Antonio for further treatment. When that did not work, his parents planned to send him to Boston for additional specialized care. The family still had roots in Bean Town, and Willie's grandmother planned to take him back herself as she had other business to tend to there as well.

Then news came that Willie's grandfather in England had died and George needed to return to settle the estate, probate his father's will, and collect his inheritance. (Unfortunately, through a tragic turn of events, he lost his inheritance to the Court of Chancery.) To raise money to defray the cost of both trips, they gathered five hundred steers and a "remuda" of horses, hired trail hands, and drove the herd to San Antonio, which was the only market in Texas at the time. The value of those steers was about seven dollars a head, delivered at the market. Willie's uncle, Joshua Robinson, his father, Bob Robinson, and his grandmother, Lucy, left with Willie on this long, dangerous trip through Indian territory. From San Antonio they planned to travel to Galveston and book passage for both Boston and England.

Although they were fortunate enough to arrive in San Antonio safely, they discovered that a cholera epidemic had struck the area. The whole city was in confusion. No one was thinking about buying anything, and most of the population was packing up and leaving town. People were dying so fast that there was no time even to dig their graves. Work crews stacked the bodies onto gruesome wagons and hauled them away to mass graves.

George Roberts Sr. had no choice but to try to wait out the pestilence, but his decision proved fatal. Everyone in the Menard County party died: his father, grandmother, and uncle. Everyone except little Willie. He was found in a wagon alongside his dead family members. Someone took him to a central station, where strangers and lost children were identified. His father, uncle, and grandmother were buried, but no one knows where. And no one knows what happened to the five hundred cattle or the money they brought at market. Providence smiled on Willie, however, as he was identified by a German from Fredericksburg named Doebler. He had seen Willie at the Nimitz Hotel in Fredericksburg when Lucy and Lucy Ann had previously taken him to San Antonio for treatments.

Doebler was then allowed by the authorities to take Willie to Fredericksburg. On the third night of their trip, they camped at a spring on the side of a long string of steep hills about twenty miles below Fredericksburg. Doebler succumbed to cholera himself, and again little Willie was to be found alone in a wagon with another dead body. Soon a passing freight hauler named Taylor discovered the boy and carried him to Fredericksburg. Taylor was on his way to Menard to build a stagecoach station at a place on the San Saba River called Peg Leg. He offered to take Willie as far as that, but when they arrived at Fredericksburg, people were so terrified of the threat of cholera that no one had the courage to take Willie off the wagon until some nuns picked him up. Some townspeople believed that Taylor showed signs of cholera himself, and they locked him in a room. He died that night.

The Nimitz family recognized Willie and sent word to his family at Fort McKavett. It was the first the family had heard of the tragedy. They saw to Willie's safe return, but all their cattle were now gone and the family was almost destitute. They never fully recovered from this terrible ordeal.

Still, the family had to operate a huge spread without a capable ranch manager, very little capital, and no fencing. Their herds strayed from the river and became easy pickings for the Indians, whose raids were still quite prevalent. There were also problems with cattle rustlers.

Large numbers of their cattle strayed from the San Saba River, ranging off onto the tributaries of the Concho and Llano Rivers. They were never found or recovered.

At one point the family needed funds, and Jack Livingston was asked to round up a bunch of the Robinson cattle and take them to a northern market. He arranged for a packhorse to carry his food and bedding for the drive and took off on the trip through Kickapoo Creek. He was gone for a long time. Years passed and he never returned, which made people wonder. Did Livingston take the money and run or had he been bushwhacked on the trail? Finally, his skeleton was found above Kickapoo Creek. There was a bullet hole in the back of his head, but no way to know if he was killed by Indians or outlaws. Dr. Bob Robinson identified the skull as Livingston's because all the teeth were molars, a peculiarity of only Jack Livingston. None of the cattle were ever found.

When Willie was six years old, his mother married Arthur Striegler, a Dane. As he got older, Willie gained some strength and partial use of his right leg. However, for the longest time, he could not stand without a crutch. His legs might have been weak, but from the waist up, his body continued to develop and was stronger than just about any man around. He later learned to walk with a cane.

As Willie grew up, he fell in love with horses. By the age of seventeen, having become pretty handy with a rope, Willie became an accomplished rider and one of the best horse trainers in the country. He was also a pretty good judge of cattle. Riding and breaking horses was not the same for him—a man with virtually no legs—so he had to make a special saddle for breaking horses. He invented a way to tie back the saddle's sweat leathers and sew on split boot tops on each side in just the right places. He could get his knees in the boot tops and ride pretty well. If he could just get a hand on the saddle horn, he was strong enough to throw himself into the saddle.

As a young boy, the responsibility fell to him to go out on the range and round up young, wild, range cows with baby calves and full udders. His mother milked them and sold the milk and butter to the soldiers at Fort McKavett. Willie credited his stepdad, Arthur Striegler, with making it possible for him, a cripple, to be able to make it on his own in the real world. Some neighbors might have called Arthur Striegler a bit cruel, because he always treated Willie as if he were as fit as everyone else. When Willie was thirteen, his stepfather sent him out on horseback to herd dairy cows. If any of those animals strayed, Willie would get a whipping. When Willie was sixteen years old, he was perfectly comfortable in a saddle, especially if it was one he designed to accommodate his infirmity.

One evening he came in from an unsuccessful search for a stray horse, and Striegler slapped his face. Willie hopped up to the house on his crutch, but he kept on going, right out the back door into the night. He left home and never returned.

On his own, Willie hired on as a regular ranch hand and over time worked for several cattle spreads. He neither asked for nor received special consideration because of his handicap. In later years he remembered working on roundups of herds of six thousand head and some even up to eighteen thousand that were pulled in from the vast western ranges yet unscarred by barbed wire.

Willie's first trail drive took him to Rivers, New Mexico. After working for this outfit for several months, he returned to Menard County. Just a few years later, he took a herd belonging to William Bevans, founder of the Bevans State Bank, and George Bradford, also of Menard, to the Davis Mountains in west Texas. Upon their arrival, the trail boss called for a volunteer to ride solo and take the surplus saddles and horses back to Menard. Willie was the only one willing to undertake this long, lonesome, dangerous trip. Outlaws and Indians were all around there. Throwing fear to the wind, Willie outfitted his saddle with a slab of bacon, a sack of "skillet" bread, a couple of blankets, and a canteen and started out for Menard with the surplus horses and saddles. He drove the twenty-some horses back to Menard without incident in record time.

Willie also rode on a drive from Menard to Alpine, where he was riding point when some of the cattle and riders came close to death. They crossed the Pecos River at a wide, reasonably shallow point, and about two hundred head began to spread out from the herd and ventured into a much deeper part of the river. Confused, the animals began to swim in a circle. This quickly created a whirlpool with enough force to suck the cattle on the inside of the loop right under.

Some of the other hands riding on that side of the river saw what was happening and knew that a lot of cattle were about to drown, but none of them were about to jump into what might be a watery grave for anyone who tried to rescue the animals. When the owner of the herd rode up on a high ridge above the river and saw what was happening, he headed his horse straight toward the river, pulling his clothes off as he rode. Willie saw this too, and he spurred his mount and jumped his horse into the river. When Willie steered his horse to the swirling cattle, he directed the crazy-eyed leaders of the herd out toward the far side of the river, saving about fifty head of cattle. His boss's intention was to jump into the river himself, not ride his horse into it, and Willie's selfless actions probably saved his life as well. This was a dangerous move for Willie. If he had fallen off his horse, he certainly couldn't swim his way out of that swift current with his weak legs.

Willie was quite a bronc breaker in his prime. He could, according to many witnesses, break horses that "outlawed" and had been given up on by other professional horse breakers. He gentled many wild ponies, doing all the roping and saddling himself, even with his shriveled legs. He accomplished this because of his incredible knowledge of horses and the unusually strong arms and shoulders he had developed. He was a top cowhand in those days.

About the time open ranging came to an end and folk started fencing, Willie began to acquire a sizable herd of brood mares. He filed an application for four sections of School Land in Menard County, and upon receiving the patent, he further stocked his ranch with cattle as well. In addition to this, he became interested in beekeeping. At one time he had an apiary of 320 hives. From those hives, in a good year, he could harvest and market as much as twenty thousand pounds of honey. Willie loved to work and never slowed down.

When he met Elar Parker, they soon fell in love. The two married in 1897 and had seven daughters. However, after his daughters became adults, Elar left him. She died in Truth or Consequences, New Mexico, in 1958.

Willie subsequently sold his ranch in Menard and bought another one in Pecos County, near Fort Stockton. In his old age he suffered macular degeneration and lost most of his sight. His legs also completely gave out as well. Not at all willing to give up, and since there were no wheelchairs in his day, Willie built himself a two-by-two-foot wooden box and attached some roller skates to the bottom. He could roll around the house all day after that, and his neighbors said he never stayed still for long. Even while in retirement in the San Angelo area, and with his eyesight mostly gone, he would still peel pecans for a little income. He never once complained about what had been dealt to him in life, but rather he was always inquiring about the welfare of others around him. His handicap caused him to develop significant upper body strength. His determination and ingenuity was range-hardened so that, with no legs, he was able to build a large spread and make his own way in life. He was known to be soft-spoken and the kindest person in Menard County.

He made his final home with his daughter, Mrs. A. N. Yockey, in Fort Stockton. Willie died in 1954 at the age of ninety-one, having lived a challenging and very productive life.

## McDougal's Draw

About four miles east of William Menzies's original homeplace on the river is an old stagecoach station called the Peg Leg (see the map of Menard County on pages 14–15). The way station was named after T. W. Ward, the late Texas land commissioner and friend of A. J. McCoy, the first owner of this tract of land. Ward lost his leg in a battle with Indians and subsequently had a wooden leg made. The tract was first patented (and deeded to the first owner) by Governor Elisha M. Pease in 1856. In order to encourage settlement of this hostile area, the state government granted tracts

Stagecoach holdup near Peg Leg

of land (called patents) to people who promised to build a home, dig a well, and live there to work the land. This tract is situated at a place on the San Saba that has several good crossings for both animals and wagons. In the nineteenth century this stage station was also a home, a barn, a general store, a watering hole, and a restaurant. This also happened to be a place where Indians regularly made their camps as well as an early battlefield between the Spanish explorers and the Indians.

Over the years, Peg Leg Stagecoach Station hosted many trail drive herds headed north, pioneers headed west, local settlers, and not too few Texas Rangers. The Butterfield Stage Line came through here, carrying passengers from San Antonio as well as the US mail to points west, such as Menard, Fort McKavett, and the settlements beyond. In addition to its already colorful history, robbers also took up just west of Peg Leg in the mid- and late 1880s. They had a hilltop lookout from which they could see stages approaching from a great distance to the east. Outlaws were always watching for two things: to see when the stage was coming so they could time their holdups and, most important, to see if there were any Texas Rangers on the roof of the stage, providing additional security. If they didn't see any rangers, they would usually hold up the stagecoach just west of the Peg Leg on the north side of the river. After robbing the stage, they would then ride south and cross the river a second time. To cover their getaway they would ride their horses up- or downriver a short ways. The robbers thought this made it more difficult to track them. There were several caves and a spring in their hideout area, and the place they liked best was a place just off the southeastern corner of William's ranch known as Robbers' Roost.

Robbing stages and bushwhacking lone travelers were common occurrences for several reasons. The territory was extremely large, there were little to no wire communications, the area was sparsely populated, and some areas that hadn't been burned by the Indians had dense vegetation. In addition, there were no lawmen around at that time. There *were* Texas Rangers, but they mostly came out after the fact. So everyone had to be prepared to defend themselves against all comers at all times.

About a quarter mile west of Peg Leg and east of the Menzies ranch is a place called McDougal's Draw. It was named after a rancher there, but not the same McDougal who lived near Fort McKavett. Since houses were hard to build in those days, especially if you lived alone, as McDougal did, a cave might suffice. And so McDougal lived very contentedly in a cave on the side of a high cliff just off the riverbank. He was one of the first sheepherders in the valley, and he had quite a beautiful spread. The land was so good that some unscrupulous men wanted to evict McDougal from it. They devised a plan to dress like Indians and kill the sheepherder so the blame would be on the Indians. They tried their best to make the murder look like an Indian raid, so they scalped McDougal after they had killed him. The ruse failed, however, and the murderers were found out. They might have killed McDougal, but his name lives on in that territory. Both McDougal's Hill and McDougal's Draw bear his name to this day.

## The James Sewell and James Bradbury Saga

A few years after William had arrived in Kenedy and Letha Ann's family had arrived in Sherman, a very dark day was shaping up for one of the family's future relatives. One of William's soon-to-be granddaughter's great-granddad was about to purchase his family's stake in the frontier with his life. His descendants (my cousins) still own the ranch just north of the Llano River in Kimble County that he fought to keep. The ranch also happened to be about twenty miles due south of what would be William's headquarters ranch.

Life started out well for the James H. Sewell family in Kimble County. Young James was the proud husband of Sarah, his sixteen-year-old bride. He and she came to live on the Rance Moore settlement northeast of Junction. Sarah's father was James Gossett, a seasoned Indian fighter, and two of her three brothers had been killed during the Civil War. Her third brother, Jack, gave frontier service as a Texas Ranger. Indeed, true pioneer blood flowed in Sarah's veins.

The Sewells had been married for four years and had two babies to bless their home with all the happiness and hope that children were well capable of bringing to pioneer homes. However, unbeknownst to them, the grim reaper was about to call on the Sewell household. In the fall of 1872 Jim contracted with a neighbor to cut cedar posts for him. Every day he left home for the cedar brake with his Colt pistol, his Winchester rifle, and his ax to do a day's work by himself. At dusk he would return home to the smiling faces of Sarah and their two children.

On a fateful morning Jim left his pistol behind and went to work as usual. No one ever knew why he left his pistol at home that day except that it was a heavy hunk of metal on his side, very cumbersome, and he had not needed it thus far while working in the cedar brake. That night, he didn't return home. A long, sleepless night ensued for Sarah. She tried to keep the faith but knew what had probably happened.

Jim's friend Billy Waits was summoned, and at noon the following day a search party of four found Jim's body three miles from the homeplace and brought his remains back to the fenced yard. There they wrapped him in a blanket to spare Sarah the trauma of seeing his heavily mutilated figure.

Where they found his body they saw tracks that indicated there were around a hundred warriors in the band that had attacked Jim. The Indians had probably been attracted by the noise of Jim's ax. While he was concentrating on wielding his ax, they were able to creep up undiscovered near enough to him to open fire at close range. Strangely, they failed to take his ax; bloodstains on the handle indicated that he had been wounded before dropping it. His rifle had been leaning against a nearby cedar tree, just a few feet away, but that was as good as a mile at the terrible moment of the ambush. There was evidence that Jim had made a run for his gun, but he was mortally wounded before he could reach it. A good number of lance and gunshot wounds riddled his body and told the story of his torturous death. His body had been stripped, his clothes taken, his head scalped, and his horse, saddle, and rifle were missing. Billy Waits and his friends buried Jim's body at the Bear Creek Cemetery.

As always, bad news spread like wildfire. Up and down the valley all the ranchers checked on their horses and animals. The watchword "Look out for the Indians!" rang from one settlement to the next. Immediately, seventy-five-year-old James Bradbury decided to organize a posse. He mounted his gray stallion named Possum and led nine others in pursuit of the Indian raiders. The posse included three of his boys. Some insisted that the elder Bradbury stay behind and provide security for the settlement's women; they were sequestered in the fortified Bradbury quarters. The old battle-scarred Indian fighter dismissed the idea as so much nonsense. He had never been sick a day in his life and had saved the day on many occasions with his nerves of steel and straight shooting.

As the men departed the stockade, a sense of apprehension and danger swept over the womenfolk as they waved good-bye and two pet bears chased each other and rolled about in hapless play. Ten palefaces found and followed the trail of a hundred Indian raiders! As soon as the posse detected the war party's trail, some thought that it might be a much smaller number than Billy Waits had estimated.

Teacup Mountain

Just south of Runnels Point, the trail turned sharply to the west, just north of Teacup Mountain and into the rugged foothills. The posse was in close pursuit. Just as they came near the foot of the mountains, known today as Bradbury Hills, they found fresh horse tracks. In the immediate distance they noticed smoke wafting up from a large brush thicket near the top of the mountains. Bradbury's posse began hugging the brow of the ridge so as to avoid being seen by the Indians. They were able to get within a couple hundred yards of the point where they had seen the smoke. Contact was imminent, and every man checked his weapons to be sure they were in order, locked and cocked, and that their ammunition was handy for the approaching fight. They believed the Indians were busy cooking their meal of horsemeat, busy making plans for their next raid, or just resting.

Bradbury led the way, rifle in hand, and was the first to see an Indian sentinel who had not yet spotted the ten white men in their concealed position in the brush on the hillside. The men debated the situation briefly, and Bradbury decided to open the attack regardless of how many warriors might be nearby. According to James Caviness, who described the battle to his sons, Bradbury said, "Come, let's fight 'em!"

Over the objections of several others, Bradbury quietly dismounted, laid his rifle over his saddle to get a steady support, took aim, and fired, hitting the sentinel. The latter, mortally wounded, staggered and fell. Yet as soon as the crack of Bradbury's rifle sounded, warriors in full battle array swarmed out of the live oaks from where they moments before had been feasting on horsemeat. There had to have been a hundred of them, Rance Moore said. "They were as thick as blackbirds!"

Immediately a speedy exchange of fire and a spirited firefight ensued. The Indians had superior numbers and also commanded the high ground. They raced down the mountain in hordes, firing and shouting their war cries. John Nabers had been riding a mule, since horses were in short supply, and the animal was shot right out from under him. Rance Moore's horse was shot in the shoulder, and although he became unruly, Nabers jumped up on the horse and rode behind him. Just as Nabers made his leap, an arrow whizzed past him and into the stomach of James Bradbury, who fell immediately to the ground, mortally wounded with the arrow hanging from his side. The old man's son James Bradbury Jr. remained at his father's side until an Indian came within a few feet of him. The warrior was preparing to throw his tomahawk at him while James threw his pistol at him, causing the tomahawk to be deflected. Another Indian ran to the dying Bradbury's side and grabbed his rifle, whereupon the brave son wheeled his horse around and, hovering close over the side of the horse's neck, rode quickly out of firing range of the savages.

When the surviving white men made their way back to the valley road, they encountered a southbound covered wagon. The posse asked the travelers to tell the women the news of the skirmish and to let them know that they would be arriving home sometime that night.

Just as dark fell, the posse traced its way back to the scene of the battle and found Bradbury's remains. The Indians had scalped him and cut a couple of deep gashes in the shape of a cross on his chest. Two dagger or lance cuts were made crosswise through his throat. Further examination of the area revealed that two different groups of warriors had met on mountaintop where the feasting was going on just prior to the encounter. About a dozen Indians had been raiding the valley, and the posse had been following their trail. The smaller group then rendezvoused with a much larger group from Gentry and Bear Creeks on the mountaintop.

Meanwhile, as the hours dragged on into the night, the women at the settlement were anxious but composed under the strain of the news of the clash between the posse and the war party as well as Bradbury's death. And now there was even more uncertainty. They walked the yard, carrying their babies and trying to sing them lullabies to calm their own nerves.

When the posse finally returned, a makeshift coffin was constructed, and the veteran Bradbury was buried under the branches of a live oak on the northern bank of the North Llano River, just below the site of a present-day bridge. There ended the last chapter in the life of a valiant and honorable pioneer rancher, a man whose life was dedicated to building his ranch, serving his family and his neighbors, as well as taming the wild Texas frontier.

Soon after this, a fresh grave of an Indian was discovered near the scene of the fight. It is believed to be the grave of the Indian sentinel who fell as a result of James Bradbury's first shot.

To this day, the land Bradbury gave his life for is still enjoyed by his descendants. It is the headquarters ranch of my cousins, Phillip and Lynn Jacoby. Phillip's mother, Letha Ann, was the daughter of William and Letha Ann Menzies. Every day when Phillip and Lynn leave their ranch, they see the Bradbury Hills and Tea Cup Mountain and are regularly reminded of the great price that was paid there many years ago.

## The Cattle Drives

Raising livestock in west Texas against nature, Indians, and outlaws was tough enough, but getting them safely to market was an even bigger problem. The distance, intense summer heat, blustering cold winters, and choking dust from the trail drive were additional obstacles that had to be surmounted as well. Still, these physical problems were nothing to be compared to the threat of Indian attacks and outlaw ambushes. A well-executed ambush by either group would only take a few minutes to carry out, and the whole cattle herd would be theirs. Killing the trail boss and a few of his drovers to get immediate possession of a large number of livestock wasn't a moment's concern either to Indians or outlaws in this territory.

There were very few cattle drives during the Civil War, and drives remained nonexistent after the Union army took control of the Mississippi River in the summer of 1863. This hiatus gave Texas cattle time to do what they do best: multiply. Since Texas sent more young men to the Confederacy

Texas longhorns

per capita than any other state (and many never came home), there were very few wranglers to round up the cattle and take them to market. Following the surrender, just about everyone in Texas was penniless. When the Confederacy collapsed, their currency was worthless. Much of the state's wealth was in slaves, and with emancipation the economy was virtually nonexistent. Then carpetbaggers swooped down and bought as many farms, ranches, houses, and businesses as they could. For a while, the steers on the range had absolutely no value, because they were so abundant. Yet many believed that if they could get them to markets in the North, they would bring big money. Lacking funds, very few had the capital to hire the cowpunchers or feed those men on the long trail drive, even if they could round up the steers for free on the open range. With all these things working against them, some ranchers, through sheer determination, managed to put together some drives and get started toward the railheads that would take the cattle to market. Those who succeeded were able to make a small fortune and then finance another drive and another until either their luck ran out or their health failed from the hazards of those drives.

Most of them took their herds up the Chisholm Trail, a route laid out by Jesse Chisholm, a prominent Scottish-Cherokee trader and guide who connected these Texas steers with his trading post deep into the Indian Territory (present-day Oklahoma). About ten million head of cattle were driven from Texas into markets in Kansas and Nebraska during this time.

To be manageable, large herds of five thousand and more had to be divided so there were no more than twenty-five hundred steers in a drive. A larger number would have been unmanageable due to the sheer expanse of the operation. Herding cattle to railheads was an expensive and dangerous part of the cattle business. It was not cheap hiring hands for cattle drives, because everyone knew well that driving cattle meant many weeks in the saddle from dawn to dark, constant exposure to the elements, and the possibility of Indian and outlaw attacks.

All this being so, the most important factor in driving cattle was finding a trail boss who was fearless, well liked, and generally trusted as a leader. The ramrod, the trail boss's assistant, had to have most of these same qualities as well.

It was likewise needful to get a good, dependable cook. His efficiency played an important role in not only expediting the drive but also contributing to the morale and efficiency of the entire crew. It fell to the cook

The chuck wagon, always a sight for sore eyes

# THE TEXAS PORTION PACIFIC RAILWAY

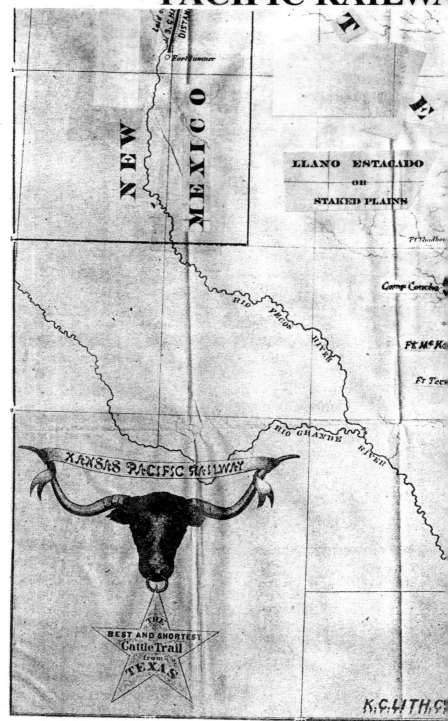

# THE 1875 KANSAS CATTLE DRIVE MAP

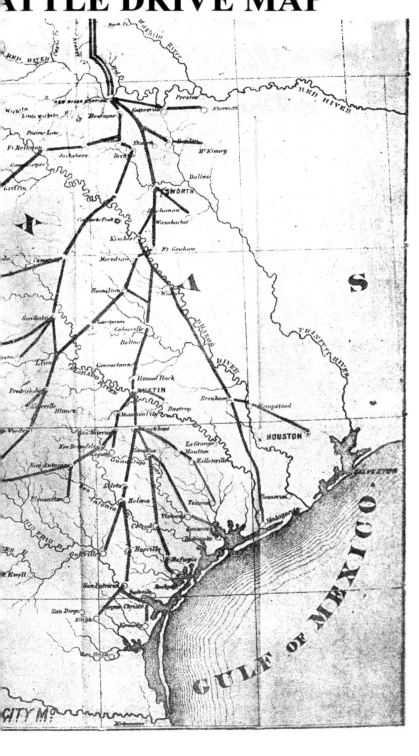

and the chuck wagon to run ahead of the herd with the scout. Their job was to be on the lookout for water, obstructions (either terrain features or someone refusing to let them pass through their land), a grassy bedground for the cattle, and a good place for the drive team to camp. It was extremely important for the cook to set up the camp, have the fire going, and prepare a meal so that everything would be ready when the cowboys and the herd arrived. A good cook was a pretty popular person on the drive team.

Cattle drives also required a horse herder or wrangler to maintain the remounts in what was called a remuda. Even the best horses only lasted half a day before tiring, and the cowboys had to be careful not to ride them into the ground. First, cowboys truly cared about their horses, and second, they certainly needed all of their horses to get the cattle herd to the railhead. On long drives, some would inevitably be injured, so having replacements was a necessity.

Two men usually rode point to lead the herd in the right direction and to look out for danger. A couple of men rode side, on the front edge of either side of the cattle herd. A number of drovers rode flank, depending on the size of the herd (more toward the rear of those who rode side), and two men rode drag to scoop up the lames and strays. These positions were rotated daily because of the horrendous dust. At night, three or four hands would ride guard on shifts, circling the cattle as the others slept. While handling these large herds, it was customary for the outriders to ride strung out, so they could just barely see the riders in front of them.

For communication and to break the monotony and keep the cattle calm, the cowboys would sing. The old cattle drive songs were no joke. (Some singers put a few of these songs on records in the late 1930s and 1940s and made a fortune—at least what was considered a fortune in their day.) Had things gotten too quiet on those drives, the slightest strange noise would spook the cattle into an always destructive and sometimes deadly stampede.

A good saddle pony was, without a doubt, the prized possession of every cowboy, especially on trail drives. A fine cutting horse, his best friend, knew how to follow a herd of cattle with a sixth sense. This horse had a slight touch of tiger in its persona, was full of energy, always gladly plunging to the mill on a round up and ever ready to dart out into the countryside to collect a lone stray. A good one can follow a stray like a swooping eagle, turning, dipping, dodging, gently biting the hide of the cow's hind quarter on every leap and turn. A good cutting horse enjoyed the fun of its work most of the time far more than its rider.

Another concern that always plagued a trail boss's mind was the prospect of running into a thunderstorm that could stampede the entire herd. Stampedes resulted in the herd getting scattered, some getting lost as strays, and a hand possibly getting trampled, gored, or killed. At the very least, they could lose precious time. Their main hope and prayer was that none of their hands would be gored or trampled in this kind of fracas.

In order to encourage cattlemen to use their rail line to ship their livestock to market, the Kansas Pacific Railway published a map and distributed it to the cattle ranchers. (See map on pages 114–15.) It closely followed the twenty-odd US Army forts spotted across the state because neither Indians nor outlaws wanted to mess with the US Cavalry. Both knew the cavalry could react rapidly to reported problems. Keep in mind, though, that a well-executed ambush required only a few minutes of gunplay for a committed foe to take out every cowboy on the drive, and steal the whole herd. Then there would be no one left to report anything, so the security these forts provided was more imagined and hoped-for than real most of the time. The forts were all built next to rivers that supplied water in a thirsty land for their livestock and a place for them to buy grub. Of course, the westernmost frontier cattle drive trail came up through Fort McKavett and Menard County. This was also the western edge of civilization, where there were still the lingering problem with Indians. All the trails led toward the Kansas Pacific Railroad on the Red River at the Texas border with the Indian Territory (that is, Oklahoma today).

## Hot-Footed Cowboy Found Dead

About the time of William Menzies's arrival in Menardville in 1887, there was a very mysterious event at the west end of the county. A young cowboy was bringing his cattle herd up from the south, where it had been pretty dry. Since there were few to no fences, most cattlemen were open ranging. This cowboy was pasturing his herd on someone else's land, and this particular landowner had a number of adult sons. Someone later found the body of the young cowboy, drowned in a dirt water tank, and his horse was also found nearby with his saddle still on. When they took off the dead man's boots, they found one of his soles had been seared with a branding iron. The landowner mysteriously turned up with a bill of sale for the cowboy's cattle, but no one ever found the money the cowboy was supposed to have received in exchange for his cattle. Most people believed that somebody tortured the young cowboy until he signed the bill of sale and then drowned him in the tank, but no one could ever prove it or find out who did it. Oddly, the matter was never investigated by the sheriff.

## Cross-Country and Dead

Just northeast of William's homeplace is the road he took to Brownwood for his lumber purchases. On that trail was a place where another rancher and an itinerate cowboy got into an argument. The cowboy was bringing his horses across the rancher's land and was about to go through his gate to leave. Evidently, the landowner for several reasons didn't like it. There were very few main roads, and everyone had to go through someone else's property to get anywhere. This rancher had words with the man driving his herd up through his land, and then he passed by the cowboy's herd headed toward his house. The cowboy rode around his horses to get to the gate the owner had just come through so he could open it, drive his herd out, and leave. The landowner had ridden off only a short way in his wagon, and the more he thought about this man driving a herd through his range and the words they exchanged, the angrier he became. So he stopped his wagon, pulled his rifle out from under the seat, took aim, and shot the cowboy dead on the spot as he was opening the gate. Then he went back and pulled the dead cowboy's saddle off, loosed his horse and his herd in his pasture, and kept them. Nobody knows what happened to the dead man or his saddle, and nothing was ever done about it.

## Cattle Feuds

Barbed wire was invented in the mid-nineteenth century, and in 1874 Joseph Glidden received the first US patent for this newfangled fencing wire. Still, it took some time for the product to penetrate the market due to the high cost of fencing relative to the dollars land was producing at the time. Without fences, however, cattle rustling and sheep stealing were rife. People did not take kindly to someone stealing their livestock, and without fences, strays mixed in with the wrong herds all the time. Explaining how you accidentally had someone else's stray in your herd required the wisdom of Solomon to adjudicate, and precious few ranchers had that

Pioneer Days

kind of wisdom or patience, and even fewer were willing to listen to explanations. Tempers flared.

There were few fences of any kind when William arrived in Menard County. Wire fences with cedar posts were expensive and difficult to build by hand on rocky west Texas land. If you couldn't find enough cedar posts on your property to do the job, all the posts and wire had to be hauled in by wagon. William's Gap ranch wasn't fenced until 1913, when he fenced it with a net wire that was advertised as "wolf-proof." It was such a heavy gauge that most of it lasted until it was replaced in 2008, ninety-five years later.

## Menard County Formed

In 1858 Menard County was formed from what was originally Bexar County, but it was not officially and formally organized until 1871, when three justices of the peace met there for the express purpose of establishing the county. Menardville became the county seat. Both the county and city were named after Col. Michel B. Menard, a very colorful leader and a great Texan. Menard was born in 1805 in Canada. He migrated south, and after spending a number of years with the Shawnee, he arrived in Texas at the age of fourteen. Menard was first employed by John J. Astor's fur-trading company, and then he became involved in land speculation with Juan Seguin, a citizen of Mexico. At that time, land was granted only to Mexican-born citizens. When the war for independence broke out, Seguin joined the Texans and fought for independence under Houston at San Jacinto. Colonel Menard was not only happy to join the fledgling Texas army, but he was also one of the brave signers of the Texas Declaration of Independence. He risked everything to help Texas break out from under the iron fist of Mexican rule. Menard distinguished himself by fighting to the finish in the war for Texas independence. He was faithful until the final great victory was won at San Jacinto.

## Primitive Living

During the late 1800s houses on the west Texas frontier were primitive. There were no screens on their windows—if they had windows. No electricity, no telephone, no refrigerator, no washing machine, and no air

The Menzies homeplace as it looks today

conditioning. No doubt they lacked many things that their kin enjoyed back east, but it didn't really matter to them. They had freedom and the thrill of forging a new life with great opportunity, building ranches, and growing their families.

It wasn't until about 1916 that pickups and cars were first owned by folk in Menard County, but even then only a few had such luxuries. Everything had to be brought from town to the ranch by a horse-drawn buckboard or wagon. Lumber had to be sawed by hand on site or hauled in very expensively and painstakingly over bumpy, rocky, rutted, winding, dirt roads from Brownwood. There were no chain saws, mechanical log splitters, jackhammers, front-end loaders, electric circular saws, drills, plus money was scarce.

They lived in an agrarian society, so people bartered, trading so many sheep for so many goats, pigs, or eggs. Building materials were expensive. Constructing a house or a barn was a Herculean task physically, financially, and logistically, and William was both destined and determined to build the many structures he needed on his ranches. He was an excellent carpenter and mason, and because of that and his construction experience, he was blessed not to have to find an architect or a builder whenever he wanted to construct something. He could design whatever he needed, lay the foundation himself, and erect just about anything he took a notion to whenever and wherever he wanted.

After the original homeplace on the river was flooded in 1899, William moved the house up the hill and expanded it significantly. The final house had a big entrance hall, a parlor, a family room with a large fireplace, a dining room, a kitchen, and two large bedrooms. Later he remodeled the house, adding a second floor with a huge room for all the boys. Even later he added a bathroom with indoor plumbing (it was all outdoor privies until then) and a huge, wide wraparound porch on the front of the house with a baluster and two small porches in back.

## The Saloons of Menardville

Certainly saloons played a colorful part in Menardville's early history. Although dances were popular, they were always held in schoolhouses, dance halls, and later in hotels. Other than the dances, church life, and occasional horse races down San Saba Street, there was not a lot of entertainment in town. There was, however, a little guitar pickin' and singin' in the saloons, but these establishments were frequented by a clientele primarily interested in gambling, drinking, and "shooting the bull."

You wouldn't say that shootouts in the street were common, but a bunch of them happened over the years. At least one man was stabbed to death. Raymond Walston, William's son-in-law, was across the street from the Legal Tender Saloon, probably at Luckenbach Hardware, where most of the family did business, and witnessed a man killed in the street. The dead man was one of two of the saloon's patrons who decided to settle a disagreement with their pistols. That particular case created such a controversy in Menardville that the judge moved the trial out of the county to New Braunfels. Raymond was subpoenaed for the case and sequestered in a motel there, but he never testified because the case was settled before his testimony was needed.

Asa Ellis operated one of the first saloons in 1876 in a building on the north side of upper San Saba Street. Just afterward, Dan Murchison opened a bar in the same area and called it the Road to Ruin Saloon. It lived up to its name very well. William Johnson built the Old Rock Saloon in 1887. There are several reliable accounts of drunken cowboys riding their horses into the saloon through the swinging doors and demanding drinks be served to them and their mounts.

At some point the Iron Clad Saloon opened across from the Australian Hotel and was run by a man named Saunders. The Gay Brothers

Saloon was built next to the Bevans State Bank and later changed its name to the Legal Tender Saloon. In 1891 the Rock Saloon was bought by Bud Westbrook and Will Callan when William Johnson, the proprietor, died. They moved it across the street and a little farther west and renamed it the Cottonwood Saloon. It went out of business in 1918.

One of the most unusual barkeeps Menardville ever saw was a fella named Jess Slaughter. He worked at the Legal Tender Saloon for a number of years. Although he stood well over six feet and was one of the strongest men ever known to live in the county. He was, amazingly, also one of the kindest and gentlest people anyone would ever want to meet as well. When all the saloons went out of business, old Jess got a job at the Wool Warehouse and worked for Allen Murchison. He went on call to the various ranches and would lift one-hundred- to two-hundred-pound bags of wool like they were nothing. Jess would also pick up mohair sacks at three to five hundred pounds a sack and get them on truck beds all by himself.

## The Railroad Comes to Town

Without controversy, the biggest happening in Menard occurred on February 10, 1911, the day the railroad came to town. Heretofore, cattlemen had only dreamed of a time when their town would have rail

The first train in Menard

*Pioneer Days* 123

San Saba Street and the Bevans State Bank in 1911.

service. This called for a huge celebration to commemorate the beginning of a more prosperous and progressive era. People came from miles around, and the festivities were many and great. It was such a progressive move for the town that it also precipitated shortening the name of the town from Menardville to Menard. Therefore, on February 16, 1911, by consent of the townspeople and by order of the commissioner's court, the name of the town was officially changed.

When it was first announced that the railroad was planning to come to town, there was so much excitement that the townspeople donated the land for the rail right-of-way, the stock pens, and the railroad station. They even built the station for free. Up until that time everyone in Menard had to fetch their freight from Brady, which was about thirty miles away. Other than going to the railroad station in Brady, freight had to be shipped by "freighters" (men who drove big wagons with horse teams). This was expensive, slow, and dangerous—there were still bushwhackers to worry about—and the freighters were also fairly unreliable. The railroad, however, was dependable and secure. Trains kept the packages dry and were much more consistent with their cartage.

Getting shaved at the barber shop to be ready for the celebration of the railroad line opening.

The great February 10, 1911, picnic on the river

*Pioneer Days*

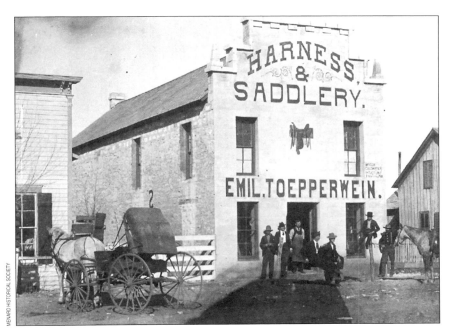

The coming of the railroad did not cripple the business of the saddlery shop.

The Menard railroad station as it looks today.

The freighters lost a lot of business to the railroad.

Adolph Beyer's blacksmith shop

# Pioneer Days

Local dignitaries easily transitioned to cars for parades.

Trains helped ranchers by hauling livestock to market, saving them from the long, hot, dusty cattle and sheep drives. This was something to celebrate all by itself. Ranchers almost always had to accompany their herds personally when driving them to market if they wanted to be sure that any or all of them made the trip safely. Another liability to be considered was the valuable weight the drive took off of the animals (called shrinkage) before they arrived at markets to be sold.

The coming of the railroad was one of the greatest days the county ever knew! Menard had never seen a greater celebration. The merriment included a parade, stump speeches by dignitaries, and a huge barbecue under the pecan trees on the San Saba River. The railroad was a major blessing to Menard for a number of years.

The coming of semitrailers, however, brought an end to the heyday of the railroad. As the rail service slowed, the station in Menard finally closed in 1972. Trucks with powerful diesel engines had the advantage, because they could pick up a whole herd of animals, any product at a manufacturing plant, or a small package at your business and take it to the recipient's front door. This eliminated the need for someone to receive the goods at a railhead destination, offload them in a warehouse, and then haul them to the purchaser's home or place of business. Trucks were also far more efficient and less expensive, accommodating smaller packages and loads as well.

But the railroad had its day. Rail service has diminished somewhat, but it has regained a lot of ground in recent decades, although not in Menard. With international shipping containers and because of the rail's efficiency and the increasingly high-cost fuel, railroads still enjoy a large place in the overall economy and transportation sector today.

## Kitchens School

Getting a good education was always greatly emphasized by William and Letha Ann. All the Menzies kids walked three miles every day to Kitchens School and carried their lunch, which was usually biscuits with butter and jelly and hard-boiled eggs. It was quite the thing in those days to try to imitate Teddy Roosevelt's Rough Riders. The Kitchens School kids prided themselves as Kitchens Rough Riders and rode their horses in the parade when the railroad came to town.

On the way home from school, the Menzies kids would gather vegetables from their garden and cross the San Saba on a foot log. They also

Kitchens School, Class of 1910

*Pioneer Days* 129

would check the trotlines they had set out permanently along the river so they could bring the fish home for the evening meal.

All the kids, including Walter (the youngest), would be sent out to the garden at the appropriate time of year to do the planting. One day when they were planting watermelons, Walter tried to keep up with the big kids, but after a while he ran out of steam. So he dug a hole and dumped all the seeds he had left and covered them up. Then he curled up under a shade tree and went to sleep. The other kids woke him up when it was time to go home. It looked like he had gotten away with one—until the seeds germinated and came up. Granddad found about fifty watermelons growing out of one hole, and he knew exactly what had happened. Somebody might have gotten their hide tanned, but we don't really know for sure.

After elementary school, the older girls—Agnes and Letha Ann—went to "normal school" and took the state examination for teachers' certificates. Back then, small neighborhood schools dotted the county, as there were no school buses in those days. Kids had to be within walking distance of their school, so that meant there had to be a lot of schools. The two sisters taught in a number of county schools like these. The boys, Alex and Max, were football inspired and graduated from high school in Menard. They

Kitchens School Rough Riders

attended John Tarleton State College and played football there. At that time, all the men at Tarleton were required to be members of the corps of cadets and to wear the uniform.

To many pioneers, a good education was considered a foot in the door of opportunity. William and Letha Ann were committed to helping their children and grandchildren in many ways, and one of these was helping them get an education. Several of their children earned college degrees, and of their nineteen grandchildren, fourteen graduated from college and received degrees.

After my grandfather George died when my dad was three, my dad had to work while he attended elementary school and high school. Most years he had to miss the first month or so of school because he needed to work in the fields to put food on the table for the family. He also spent many hours walking San Saba Street and asking the men, "Shine your shoes, sir?" He and his brother G. C. also worked and paid his their own way through two years at Tarleton State College. They lived at the farm and were called "farm rats." My dad drove the bus that took all the guys at the farm back to the college every day. All during college, Dad and G. C.

Anne Crawford (later Menzies) in a Model T in 1914.

*Pioneer Days*

Tarleton students known as "Mr. Hale's Hungry Boys" per the inscription

worked for William at the headquarters ranch during the summers and dearly loved it. The pay was great. They worked for twenty dollars a month and their "keep" (namely, room, board, and laundry).

After graduating from Tarleton, Dad asked William to loan him $250 to help him finish his last two years at Texas A&M. Dad lived dirt cheap back then with about thirty other guys in what was called the Tarleton Project House. Other than paying ten to fifteen dollars a month for the house mother to buy food and cook it, that is what it cost him to live. He was an off-campus member of the cadet corps and had a part-time job in the science laboratories. That $250 Granddad loaned him was all the money he had (other than the income from his part-time job), and believe it or not, it carried him all the way to graduation.

In May 1940, when Dad graduated from Texas A&M, he was flat broke. The country was still climbing out of the Depression, and the help Granddad gave him made his college education possible. Someone might say, "Two hundred fifty dollars is not much help." Well, that was a lot of money back then—and Dad told me how much he appreciated it and paid it back. Just a little help at the right time in a person's life can make a tremendous difference. That $250 loan allowed my dad to have a professional career in life as a pilot and squadron commander in the air force, and he certainly wouldn't have been able to do it without Granddad's help at that critical juncture. He was always grateful for not only that but all the other things William had done for him. William and Letha Ann treated him pretty much like a son.

## Diligence Required

All the activities of such a large farming, ranching, and harvesting operation required organizational ability and a lot of common sense, diligence, and discipline. There was always something to do. On rainy days, the boys mended sacks, rolled twine, or shucked corn. They learned what work looked like and soon found out that it had to be done quickly, efficiently, and with excellence. This might sound like a grim, task-driven family, but it was actually far from it. Lighthearted humor and joy ran throughout the entire family.

Tarleton boys in front of Academic Hall

The boys found excitement everywhere, including riding calves and breaking horses. When you have 150 mares foaling all the time, breaking horses is just a normal part of everyday life. Although they weren't the only ones who did it, Bill and Max were appointed by Granddad with the responsibility for breakin all the broncs. It was certainly fun and definitely an exciting challenge, but it was also very dangerous. Fact is, sometimes you break the horse and sometimes it breaks you. Horses have been known to zig when you thought they were going to zag. They are awfully talented at running their rider into the corral fence, throwing him into it, and then stomping him in the dirt to boot. On other days, they are likely to just leap up and fall down backward under all the stress of trying to buck the rider off their back and then roll over on him. Uncle Max broke his right knee three times from horses falling on him. My dad broke his collarbone while riding a brahma bull, and my cousin Carl was thrown off and hit his head while breaking a bronc. He suffered a concussion and had double vision for three months. At the time he didn't know if he would ever see well again. In addition to breaking broncs, the boys also rode horses and logs in the floodwaters of the San Saba.

As little boys were prone to do, they rolled their wagons all over the countryside, directed only by their imaginations. The boys would go off to the barns whenever their parents needed to talk about "important things" and to knock down hornets' nests with rocks and dirt clods. Dodging the angry insects was great fun! They hunted and fished to their heart's content, and the whole family was also very active in community life. There were church socials, school parties and plays, and ice cream suppers at the homeplace. There was enough fun that the kids never resented the work.

In the midst of all this activity, William and Letha Ann maintained a kind, cheerful attitude and made certain that they were never overly critical of their children.

## Family Get-Togethers

There were innumerable wonderful family get-togethers. The Menzies family was not rich in terms of dollars and cents, and the Depression no doubt had a pronounced effect on the family's finances. Indeed, there were additional demands created by having to provide for a large family, buying as well as improving and maintaining

Left to right, Ray Jacoby, Perry Menzies, Roy Jacoby, and G. C. Menzies

Left to right: Alex, Max, William, Letha Ann, Walter, Pearl, G. C., Ella, and Elsie Dietz, ca. 1915.

Left to right: Max, Hub Jackson, and Alex on a snowy day.

Bill and Cora Chastain

Neighbors from east Menard who came to visit.

Touring the country

large tracts of land, and buying, feeding, and caring for so many large herds of animals. It was difficult sometimes to make ends meet, but this family was rich in friendships and a happy family life. They also had many lifelong bonds with other family members, neighbors, and the many customers of their copious business ventures.

In order for all this to continue on a long-term basis, the family first had to be rich in their devotion to God. They not only knew love, humility, and acceptance but also knew about forgiveness and how to be forgiven.

They had family all around them, and over time, neighbors soon become like family too. A couple of Letha Ann's brothers and their families lived right across the river. Jim and his wife, Clara, had thirteen children, and her brother Will and his wife had two. Perry and his wife lived in Denison and came to visit William and Letha Ann quite often. Perry had a farm there and, among other things, raised Jack Russell terriers. On one of his visits to the Menzies ranch, he gave one of the dogs to Letha Ann, and the little terrier soon became her pride and joy. Later she became a breeder herself. Her brother Bill and his wife, Cora, who was originally from Hext (a town about five miles east of the headquarters

# Pioneer Days

When neighbors visited, they wore their Sunday best, including bonnets and hats. And they always posed for the camera.

ranch), took their inheritance when their parents' estate was settled and moved to California. There they invested in an orange grove and other farming and rental property interests that did very well. They never forgot the family and came to visit every year over the 1920s, 1930s, and 1940s. To do so, because of the desert heat, they had to drive mostly at night in their Model T. It was, however, always worth the trip. Bill was a pleasant fella, full of fun, always playing tricks on everyone, and telling jokes. Other than Bill Chastain, the rest of the twelve siblings lived in other parts of Texas, mostly around Fort Worth. Even when many of the kids left Menard, there were still many kissin' cousins who continued to live in the vicinity, and there was much visiting.

## Texas Hunting Trips

William and his brothers, Alex and George, maintained a close relationship throughout their lives. Alex and George lived in New York, but regularly Alex and occasionally George would ride the rail to Texas, getting away from their interests in the Big Apple to visit the family on the ranch. They loved to hunt with William, fellowship with the family, and generally enjoy the Texas countryside. They jumped on the rail like jet-setters

Letha, Letha Ann, Walter, Pearl, Agnes, and Uncle Alex's dog, Sport.

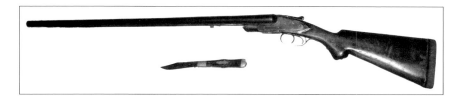

Alex's shotgun and knife

visit far-flung tourist spots today. It was quite the thing to do, and some of the railcars were quite luxurious. Most of the coaches were not plush, but they offered the passengers a wonderful view of the countryside. The dining cars were particularly well appointed. The tables had white cotton tablecloths and napkins along with fine china and silverware, and liveried waiters attended to the diners. The sleeping cars manufactured by the Pullman Company were very comfortable, with curtains covering the bunks and occupants.

One lady from Menard who hadn't traveled by train before remarked after her first ride that she went to sleep one night and left her shoes on the floor. Before falling off to sleep, she became concerned about them, so she glanced down to check a little while later and saw that they were gone. She was sure they had been stolen and went to sleep worrying about what she was going to do without any shoes the next day. The next morning, however, when she awakened, there sat her shoes right where she had left them, except for one thing. Someone had shined them for her. It was just another service, which was standard when riding the rail.

Alex and George shared their own hunting dog, Sport, which they left at the Menzies ranch. Alex also had a double-barreled twelve-gauge shotgun and a huge knife that he left with William. My cousin Scotty still owns that shotgun and knife.

The whole Menzies family, women included, of course, have always been interested in guns and hunting. Ann Crawford Menzies, the wife of William's son Bill, had the neatest ten-gauge shotgun, called the long Tom. Ann was a very slender, handsome lady, and few people could imagine her owning a shotgun like this, but it was her weapon of choice, and she was quite a hunter. The nickname for this shotgun was spot-on, because the barrel seemed like it was a mile long. When someone fired it, the thing sounded like a cannon, and it kicked like a mule. Ann was also

The result of a day's turkey hunt

an extremely hard worker, and it is said that she followed Bill's every step in the cow pens.

## Rio Grande Wild Turkey

In the late 1800s, Menard County teemed with wildlife, especially in the San Saba River Valley. Quite prevalent to this day, but especially in the early 1900s, were the wild turkey. Native to this area of Texas, the Rio Grande wild turkey thrives in this semiarid country. They take well to the protection of the bushy, scrub rangeland. At full maturity, this meaty bird reaches a height of four feet and can weigh up to twenty-four pounds. These birds are pale gray to copper in color.

Originally, when only Indians roamed the land, turkeys existed in the millions. But with the influx of Spaniards, Mexicans, and Anglos, the fowl were depleted by 1920 to critically low numbers. In the 1930s a formal program of trapping and transplanting had to be initiated to get them back into their ancestral range and their numbers up to a proper, sustainable level. Today, just by limiting the hunting season, they are plentiful again.

Double-barreled shotgun

## Mourning Dove

Another bird hunted widely in this area from early times is the mourning dove. They are both abundant and widespread in Texas. They were named mourning doves because their call is that of a mournful bird, but for all who live here, the call is one of the sweetest, most welcoming sounds of the wild. They are primarily a big-meat bird, weighing three to six ounces. These doves are gray in color with a small head and some black spots on their wings. They have a long, pointed tail, and their wingspan is from fifteen to eighteen inches, which gives a whistle in flight. The female builds and weaves the nest from sticks and nesting material provided by the male. The mourning dove is quite prolific in that the female almost invariably will lay two-egg clutches at a time and can have up to five or six clutches a year.

White wing doves have also been around the Texas countryside for a long time, but they concentrate more around cities. These birds are bigger and will grow to about the size of pigeons. The ring neck dove also seen in the wilds today was brought here as a domestic bird, but it escaped into the wild and remains quite abundant for hunting.

## White-Tailed Deer

Hunted more than anything in William's day and even today is the swift and beautiful white-tail deer indigenous to this part of the country. They come in many varieties and have always been particularly well suited to the arid Texas hill country. Deer are cud chewers, or ruminants, and have a four-chambered stomach. The stags, harts, or bucks as the males

Left to right, Raymond Walston (holding Willie Lee), Walter Scott Robinson, and Alex Menzies. Alex is holding a Model 1894 Winchester 30/30 long barrel, lever action, Hunter's model that originally belonged to William Menzies.

are called in Texas, grow to a good size and have a huge rack of antlers if they are fortunate enough to reach old age. Some have grown to weigh as much as two hundred pounds. The white-tailed deer is one of the largest species of deer in the world, but can easily hide in the bushy live oaks and mesquites to proliferate quite well. Unlike sheep and cows that consume large quantities of low-grade, fibrous food, deer are selective eaters, preferring tender leaves, fungi, lichens, fresh grass, and easily digestible shoots. They provide a significant and tasty source of meat for the native ranchers as well as the hunters who come from all over the country to hunt them today.

## Javelina

In quite considerable numbers the wild musk hogs, or javelina as they are called by their Spanish name, could be found in good numbers in Menard County during the open-range era. They are rarely found alone, as they travel in herds of fifty and up to a hundred. These hogs are quite ugly, and all of them seem to have a bent for being furious. Fully grown, they usually weigh only fifty to seventy-five pounds, but when closely pressed in the wild, with the entire herd in battle array and indignation, with all their jaw popping, boffing, and foot stomping, they can be pretty intimidating. If a man's gun ever jammed, being chased by a whole pack of boffing javelina, he would need no instruction to look for the nearest tree to climb. The animal's sharp tusks, turned upward, are definitely something to fear. They know exactly how to hook their prey with those tusks as well as how to use them as knives. Although not much for food consumption, these animals could wreak havoc on a rancher's garden in a hurry.

Pat Everett (great-grandson of William Menzies) currently manages the headquarters ranch and is a guide for hunters

## Timber Wolves

Another animal that ranchers were always glad to hunt were the timber wolves. Wolves loved to feast on smaller animals, like chickens, rabbits, and ducks, but they were forever up to making a meal of small sheep and goats. When they travel in packs, wolves could hamstring a full-sized steer or, if alone, kill a midsized calf. For this reason, wolves and their cousins the coyotes were particularly hated and hunted by the ranchers.

## Panthers

Easily making a place for itself and quite plentiful in this country, but rarely making it to the ranch house dinner table, was the panther, or mountain lion. In early ranching days, few nights passed without hearing the scream of panthers. It was catlike but about a hundred times as loud. Despite all the damage they did to burros and colts, the panthers' prey of choice was sheep and goats. Considering all the havoc they wreaked on small, helpless sheep and goats, you would think they would have been completely eradicated, especially by now. Not so. They are far too wily. The panther is still spotted around Menard today. Similarly visible is his cousin the bobcat.

My Uncle Bill and Aunt Ann trapped and killed them when they were ranching west of Menard. We happened to visit them before they came back to the homeplace in the early 1950s. They had caught a bobcat at the time and were keeping it in a cage on their place to collect its urine in order to bait traps with its scent. My parents told me not to put my fingers in the cage. I hate to tell you this, but most all the time, you could count on me at that young age to do whatever I was told not to do. But on this rare occasion, after hearing that bobcat hiss, flash his teeth, and paw the screen on that cage in lightning fashion, even I could see the wisdom in not trying to play with this animal.

This great and beautiful land was a rendezvous for all kinds of wild game. Rabbits and squirrels are still as plentiful as in days gone by, and they remain a bountiful offering on many ranchers' table. There is always plenty of animal life in the wilds of Menard. Today, there is a plethora of possums, armadillos, and rattlesnakes, as well as the occasional porcupine roaming the countryside. Hunters, especially back in William's early days in Menard, were rarely, if ever, disappointed with the produce of their hunt. Many times William would send his son Max to the river bottom with two shotgun shells and expect him to bring back enough game for the family's supper.

## Social Life

At the core of what kept the family together and what guided them in their most trying times was their spiritual compass, their faith in God. The church was a big part of family life. Church dinners were the social pleasure of that era. The whole Menzies family went to play parties at their neighbor's homes. We don't have play parties today, but they were especially popular with teenagers back then. At these kind of get-togethers they would pantomime, act out skits, sing songs, chit-chat, and play cards. This kind of get-together was also approved by the families that did not allow their kids to go to dances. The young children attending with their parents would play circle games while the older folk talked or played

*Pioneer Days*

dominos or forty-two. In the wintertime, they bedded the babies down on beds stacked high with coats and wraps.

## Ice Cream Socials

In the summertime, the Menzies household customarily hosted big ice cream socials, which was just one more benefit to having their own dairy. All this was long before ice cream could be bought in the drugstores and the markets of Menard. Intricate plans had to be made for the ice, which was especially tricky in west Texas during the stifling hot summer. Arrangements were made with the mailman, because he drove his mail hack to Mason each night and came back to Menard the next day. Granddad had Max take a roll of quilts and some money to a mailman named Worman usually on a Friday, when he stopped on his regular route at Kitchens School. The next morning, after the mailman had picked up the mail from the post office, he would circle by the ice plant to get a three-hundred-pound block of ice. He would wrap it in the blankets and tie it securely so he could first deliver the mail in Menard and then be at Kitchens School around two o'clock. Max and the mailman would transfer the ice to our family hack, and Max would head for the homeplace.

Ice plant

Ice cream freezer

Later that afternoon, neighbors would pour through the ranch gates, bringing cakes, ice cream freezers, salt, and the ingredients for their favorite ice cream. About this time, Letha Ann and her daughters would have baked a bunch of their own cakes, and the smell of fresh-baked goods wafted through the light summer breezes. William would supply the ice and the milk from his dairy for all the ice cream churns. This was long before anyone in the county knew anything about refrigeration, so ice cream was quite a treat.

William and Letha Ann loved to give, just to be giving. But like Ed Beachamp once said, "There is a destiny that makes us brothers, no man goes alone. For what you put in the lives of others always comes back into your own." The love they shared was always returned, making the Menzies ranch the meeting place of the east Menard community.

It was no small gathering when the neighbors from five or six miles around arrived with all their kids. Indeed it was quite a happening! Such a social occasion often attracted people from as far as twenty-five miles away. Of course, a lot of young men came to look over the prospects as well as to enjoy the ice cream, cake, and trimmings.

## Training the Children

William and Letha Ann devoted much of their time to training their children. Accepting responsibility and learning to do the right thing were high on the family agenda. William modeled courage to his children, and they were infused with it. The ancient Chinese proverb teaches, "I heard and I forgot. I saw and I remembered. I did and I understood." William knew

Pioneer Days

courage could only be learned when there is an opportunity to take risks and succeed, along with the pressure of the possibility of failure and its painful consequences. That is why they trusted their children with large tasks at unusually early ages. Opportunity is a great thing, but it only comes along with risks and hard work. The reason most people can't find opportunity in life is because it always comes cleverly disguised as hard work, risks, and responsibility.

On one occasion, Bill and Max were trusted to make an overnight trip in the family hack to San Angelo, a trip of almost seventy miles. They were to meet their sister Agnes and bring her home. She had been away at a normal school, studying for her teaching certificate. She was also very much in a hurry to get home and see Raymond Walston, whom she later married. They had already been keeping company, had gotten past the sparkin' stage (when the chemistry definitely seems to be a match), and

Young Max and his horse

had gotten into the spoonin' stage (looking admiringly at each other) by then. William had cautioned the boys, who were about fifteen at the time, to camp out at Kickapoo Creek but not get too close to any of the freighters. Most of the freighters were a pretty rough sort. William feared that their horses would get mixed up with the freighters' and trouble might ensue. It was nothing for William to send Max alone with twenty mares and a stallion from the headquarters ranch east of Menard all the way to the Gap ranch in the north end of the county, about twenty-six miles away.

Some lessons the boys had to learn on their own, and ranch life and nature were always right there, ready to teach them. Until the advent of the horseless carriage just after the turn of the twentieth century, a young man and his horse were just about inseparable. The boys were constantly training to manage the ranch, the livestock, and to subdue natural predators. Of course, only when a man got married and began moving his family around in a horse- or mule-drawn wagon would things really begin to change. Until then, for a young man, his horse was his daily companion and his sole mode of transportation. Horses were so critical to cowboys on the frontier that the penalty for horse theft was hanging. Hours and hours were spent by young men in training their mounts.

One thing the Menzies boys loved to do was to train their horses to leap off the riverbanks while they were riding on their backs. Training a horse in this could have been a necessity in order to catch up with an animal or something like a stray cow, but it was probably more for fun than anything else. They mostly did this at Grassy Point, which is above the homeplace, where there was a gradually increasing bank that overlooked a turn in the river. The turn naturally formed a large pool of water, which is still there today. There is a huge tree on the bank, and a rope is always tied to one its thick branches for kids to swing out and flip into the river while swimming.

The boys would train their horses first to leap off the bank from the lower levels first. This allowed the horses to gain confidence incrementally. They would trail each other and leap in, one right after the other (like the Rough Riders), swim to the far side, circle around, and do it again. What fun! Then they would gradually go up the bank higher and higher, until they got to the top, from which there was about a fifteen-foot cliff. What excitement! They were all doing this very well one day,

*Pioneer Days*

The bluff at Grassy Point on the San Saba River

leaping off from the top, when one of them almost drowned. The boys decided on their own, then and there, to discontinue this practice for good. They figured some things might be too dangerous after all.

## Chapter 8

# A Progressive Rancher

## Branding Livestock

In the early days in Texas, horse and cattle rustling was a huge problem, especially when the ranges were wide open. Even after fencing went up, rustling continued to be a problem for ranchers whose horses and cows were scattered across several thousand acres. So in 1888 William registered his first brand X, but he did his branding a little differently than others. His brand was placed so that it was visible only on the inside bulge of the animal's left hindquarter. This was a spot where few horse thieves ever thought to look, because the area was visible only when the animal walked away from you, and you had to be looking there intentionally. Unless you had a good look at it, you might think the brand was only a scar.

Branding a calf

A year later William changed his brand and registered the new one: an "open A 6." The figure 6 was a continuation of the right side of the open A, but the figure 6 didn't close, because closing it botched the flesh when the skin of the animal scabbed over and healed after being branded. Several of the original branding irons are still owned by some of the Menzies family.

## Custom Harvesting

In farming, ranching, and building, William was very progressive, preferring to live on the leading edge of the latest technology. He liked to buy the latest machinery. Some said, "He could have owned half of Menard County if he wouldn't have spent so much money on machinery." William loved machinery, and he needed everything he bought. He had a lot of country to ranch and many fields to plow, plant, harvest, and maintain. He not only had to plow and harvest his own fields, in addition to everything else he was doing, but he also operated a custom harvesting

Grandsons Duery Menzies and M. D. McWilliams brand the wall in the Kleberg Center at Texas A&M University.

Open A 6 brand.

company. William threshed grain for nearly all the farms along the San Saba River, from Hext to Fort McKavett, along some twenty miles of river bottom. He, his five boys, and a number of hired hands contracted on a percentage basis with other ranchers to provide this valuable harvesting service.

William's first machine was a mechanical thresher that was powered by a combination of sixteen mules and horses. The horses were used to guide the team, and the mules did most of the pulling. Just hitching up the team and keeping it going in the right direction was a job in and of itself. As the Industrial Revolution produced its newest equipment, a new threshing machine became available, powered by a strange contraption that jumped and burped and shot out steam. It was the steam-powered tractor. William

Sixteen-horse-and-mule threshing machine

Steam tractor

might have been the first person in Menard County to purchase one of these newfangled beasts. Although it was quite sophisticated for its day, the machine also required a lot of extras just to operate it. The contraption needed a water wagon because of the constant steam loss, and it needed a wood wagon for a ready supply of fuel. The tractor moved from field to field under its own power, but it had to be stationary when in operation because a long leather belt connected it to the actual threshing machine about thirty yards away. This all worked in conjunction with a team-drawn buck rake that gathered the crop in the field to be threshed by the thresher.

William and his sons took turns camping out at the site of the threshing. Some would stay with the machine overnight while the rest rode home for the evening. Someone had to stay with the equipment and the harvest for security. The people whose crop was being harvested had the responsibility of providing food for the Menzies men since they had to stay with their equipment until the job was done. William said he got so tired of eating red beans, buttermilk pie, and cornbread. The vittles provided were not something you would write home about. That is why it

*Pioneer Days*

Threshing cutter

Threshing machine

was such a treat to spend the night back at the house and enjoy Letha Ann's fine home cooking.

Years later gasoline-powered tractors came on the scene, and of course, William was one of the first to buy one. These machines were so much more efficient and far easier to operate. It meant there would be no more cutting, no more splitting, and no more hauling wood or water for the steam engine.

## Terracing the Land

In addition to his dairying, farming, and ranching operations, William worked tirelessly at terracing his land to better retain the topsoil, as well as increase rainwater retention and absorption in this arid land. He also planted numerous fig, peach, and pecan trees along the river bottom, which he budded to improve varieties.

## Community Irrigation Canal

One of William's greatest improvements to his land was an irrigation canal system that he built in conjunction with several other upstream neighbors. In 1892 he was instrumental in organizing what came to be called the Kitchens Irrigation and Manufacturing Corporation. Working as a catalyst

Mule team slip

Hay rake

with John Alexander, Dick Kitchens, and Jim Chastain, he helped to build a major, invaluable dam and irrigation canal system. It stretched from the Five Mile Crossing, east of Menard, all the way to his homestead farming fields, as well as to the Chastain and Jackson places.

The system consisted of a dam (called Jackson Dam) on the river and a five-mile ditch system with metal flumes across creeks and ditches that also had gates to control the water that was taken out. The irrigation canal was six feet wide at the bottom, eight feet wide at the top, and two feet deep. It was built at a two-foot slope per mile. As designed, it had a carrying capacity of thirty-five cubic feet of water per second. They also built many concrete, steel-reinforced bridges across the canal to facilitate vehicle traffic. At the time, the only vehicles were horse-drawn wagons, but all these small bridges are still there and in good condition today, after a hundred years of much heavier truck traffic. The canal was built by using a slip, some mules, and a lot of backbreaking shovel work. There were no bulldozers, jackhammers, backhoes, or front-end loaders. If any equipment like that was even in the country's big cities, there sure wasn't any in Menard County.

Field hands preparing a meal over an open fire

The shareholders in the canal corporation had to invest $2,000 to build it; that would be about $49,600 in today's dollars (according to MeasuringWorth.com). That amount of money would have been enough to buy about five to seven hundred acres of land. It is an even larger amount of money when you consider that few people had much money in rural ranching communities. In an agrarian economy, people bartered and traded for what they needed: so many chickens for a hog, and so on.

William was the first secretary of the Kitchens Irrigation and Manufacturing Corporation and later served as president. The canal company flourished. It issued a hundred shares of stock, and each participant received so many shares based, of course, on how much each person invested. In return, each share entitled the owner to so many minutes a week of irrigation time to take water from the ditch. That irrigation canal, though difficult to build in pioneer days, paid each investor hefty dividends year after year in terms of real production on their respective farms. They were all able to turn what was nearly useless land into highly productive farm acreage, and most important, their participation gave them a water supply even during droughts. The canal blessed the original investors every year for the rest of their lives as well as their children and future generations.

## Lighting Systems

The Menzies homeplace was first lit by kerosene lamps and then by carbide gas, which William piped throughout the house. Later he bought one of the first wind-powered electric generators. He erected a windmill on the hilltop just west of the homeplace and stored the electricity in thirty-some huge glass jars that held the electrolyte. These jars were kept in a concrete-walled building behind the home. Despite all its size and intricacy, the system only supplied a small amount of light for a short time; but it was quite a blessing at the time, and the family thought it was well worth it.

When the local power company began generating electric power in Menard, William was one of the first to sign up and agree to a long-term contract. He paid a monthly fee until it later evolved into an individual meter system.

# The Horseless Carriage Comes of Age

As soon as motorized vehicles were available in west Texas, William purchased one. Sometime between 1918 and 1920, he bought a Dodge. They didn't have it too long though, and for a reason few would expect.

One day William and Letha Ann were driving the car from the homeplace to the garden to gather vegetables. While they were picking, they noticed a dark storm cloud coming in quickly from the west, so they loaded things up and quickly headed back home, through the river—there was no bridge here. Soon they found that the rain upstream had changed the normally clean, clear water to a muddy brown. More important, a wall of water rose up and was coming at them so quickly they had to leave the car midriver and wade speedily through waist-deep, fast-moving muddy water and debris. Just as they made it to the riverbank, the current swept the car into the deep water hole below the crossing and flipped it completely over. They lost their new car, but at least they were blessed, once again, to escape with their lives. Later they recovered the car, cleaned it up, and sold it to someone who drove it for eleven more years.

William next purchased a Model T pickup. After he purchased it, my dad (William's grandson), was working for him over the summer. On one particular day, all of dad's uncles and cousins were working with William in the field. William had a large canteen full of drinking water for the crew. It had been set off the truck and out in the field, so the truck could be used for other purposes and the water would be readily available. Dad was only fourteen or fifteen at the time, and they were teaching him how to drive the truck. He was trying hard to look like he could do it, and he wanted so badly for the older guys to be proud of him. But while backing up, he forgot about the canteen and ran over it, crushing it beyond repair. He so wanted to be a field hand, but his name that day was mud. He told me, "Son, never drive a car in reverse an inch farther than you absolutely have to." That has always been a good piece of advice. Insurance companies have determined that 50 percent of all auto accidents happen while a vehicle is in reverse, and we only spend a fraction of a percent of our drive time going in that direction. Dad did a whole lot better flying for the air force, since none of their planes had to go in reverse.

Not long after the first pickups came out, William bought what was called a bobtail truck. It had a long, flat bed on the rear with removable side panels. This truck could haul large amounts of just about anything from lumber, fence posts, and feed to equipment. It was so stout that it

Model T pickup

could even haul large bulls. Once William used it to haul three bulls at one time out to Tom Jacoby's ranch. Motorized transportation had then come of age, and trucks facilitated William's ranching and farming operations like nothing before or since.

# World War I

While things were going well at the ranch, big trouble was on the horizon. A greater threat to our nation than our own civil war and certainly the Spanish-American War that had just concluded were the ominous clouds of World War I that blew westward from Europe. The fateful match was struck to the fuse on June 28, 1914, with the murder of Archduke Franz Ferdinand of Austria at Sarajevo. Following this, there were several different explosions across Europe. Exactly one month later, to the day, Austria-Hungary declared war on Serbia, and the rest is history. Within days all of Europe and Russia were sucked into the fray, and England joined in shortly thereafter. Despite provocations over the next two years, the sinking of the *Lusitania* on May 7, 1915, compelled President Woodrow Wilson to renounce our policy of neutrality and attempt an ill-fated effort to broker peace between the warring nations. Soon, Germany's actions made the decision for us.

On April 6, 1917, German submarines deliberately sank more US ships, and then German agents attempted to incite Mexico to initiate a war against us. At the time, Germany had its hands full with England and France, and the kaiser's advisers feared that the United States might side with the Allies, so they tried their best to exploit our problems with Mexico. German foreign minister Arthur Zimmerman dispatched a diplomatic offer to Mexico, proffering a military alliance with Germany if the United States ever went to war with either of them. They further promised to return all the land that Mexico had previously lost to the United States (specifically Texas) once they had achieved victory. The only problem with this dispatch, as providence would have it, was that the British intercepted the note and published it abroad. This incensed the United States, especially Texas. All these developments gave our president no choice but to declare war against Germany.

World events once again reached all the way to Menard County, Texas. Young men were called up to fight in foreign lands they had never heard of, and their families didn't know if they would ever see their sons again.

The draft board placed newspaper ads requesting volunteers for military service. Married men were usually deferred by the board, and mostly unmarried men were drafted. Although Raymond Walston had eyes for Agnes Menzies, he was still single. Since his brothers had already married, he volunteered to serve his country in the army. After basic and advanced training, he was sent to Europe to serve as a farrier (a person who shoes horses) on the front lines in France. There were precious few motorized cars, trucks, and tanks in World War I, so almost everything—equipment, weapons, material, and personnel—was transported by horse-drawn vehicles. Because of this and the introduction of so many advanced munitions and the high center of gravity of horses, the draft animals constantly suffered a tremendous number of burns, lacerations, and contusions. The overall ratio of horses to men was extremely high just to keep the army supplied and mobile.

Also called upon to serve his country was one of William's sons: Bill Jr. He joined the army and served his country as well.

Years later, Agnes Menzies was teaching at Kitchens School on November 21, 1918, when young Walter had to stay home sick from school. Later in the day, the Menzies family heard (probably by telephone) that an armistice had been signed between the warring nations. Sick as he was, Walter saddled his horse and rode across the river to Kitchens School to share the news. Upon hearing the great news of the Allied victory, there was much excitement. School was dismissed immediately so everyone could go home and celebrate. Not too many months later Raymond Walston and Bill Menzies returned home to the joy of the whole family.

## The Great Coyote and Wolf Roundup

Life on the ranch was never without problems. If it wasn't one thing, it was something else that required ranchers to stay alert and on their toes. Coyotes and wolves have always been a problem in west Texas, but in 1923 this problem had gotten completely out of hand. The sheep and goat herds were being decimated by these pesky beasts. Not to be defeated by anything, William picked eleven men to join him and go to all the ranchers across the whole county, recruiting members in a varmint eradication association. He devised a voluntary donation plan for all the members of five cents per head on cows and two cents per head for every

Bounty hunters going on a coyote hunt

goat and sheep to raise the money needed to pay the bounty hunters to kill the wolves and coyotes. Their eradication program proved to be very successful.

Today, we see ourselves running up on the same old stump. Our sheep population is once again on the decline. This dramatically affects both our economy and our ecosystem. Many products and by-products come from sheep hide, meat, wool, horns, hooves, and bones—bringing in an estimated $7 billion a year. Things like hand lotion, detergents, luggage, gloves, photographic film, adhesive tape, buttons, and piano keys come from these animals. Sheep grazing helps the ecosystem by preventing forest fires and eliminates toxic and noxious weeds, making a healthy habitat for other animals like elk, antelope, and deer. According to Glen Fisher, president of the American Sheep Industry Association, sheep numbers are still shrinking. At their ninety-fourth annual convention in San Angelo in August 2009, he reported that the 7.05 million head of sheep in the United States are down 3 percent from last year, and the goat population of 3.71 million is down 2 percent. Although Fisher conceded that drought and high feed costs had an effect on this decline, the main reason for the drop was increased problems with predators. Without predator management as a critical and constant part of sheep production, Fisher said, they will "eat

A bounty hunter, William, and Alex pose with a wolf that had plagued the Gap ranch for five years.

the sheep producer out of operation." A rancher said, "Coyotes kill for a living [and] there is no magical way to train a coyote to not like the wonderful taste and tenderness of lamb." Maybe it's time for William's bounty hunters to saddle up and ride again.

## The Eagle Problem

Another difficulty William faced were the ravages of bald eagles. Unlike buzzards, who feast on the dead carcasses of just about anything, eagles are opportunistic feeders. They much prefer fish, baby calves, lambs, kid goats, deer fawns, and rabbits. Over just a few weeks, a lone eagle can easily devastate a livestock herd's production. For a rancher who makes his living off of livestock and is trying to raise a family, this is a huge problem. Everyone loves the American eagle, and one might have a difficult time with ranchers having to kill them, but the problem is mostly one of perspective. People in the city get the meat they eat in very attractively wrapped cellophane and Styrofoam packages in air-conditioned stores. We have very little notion how difficult it is for a rancher to make a living. We have probably never been kicked in the shins by a cow or a horse or breathed in a bucket of dust while hefting eighty-pound bales of hay into the top loft of a hot barn in July. We likely have never sat on a ten-inch stool in the bottom of that same dusty barn for two hours twice a day, milking cows, slopping pigs, or having to wring a chicken's neck and then pluck, gut, and cook it.

Eagles are extremely crafty, deadly birds of prey, very damaging to young livestock and difficult to discourage. They wait patiently in treetops, scanning for their prey. Then they swoop down to the ground, grab their prey, and then go dipping, drifting, and dodging between low trees to find a high tree in which to roost and begin their feast.

Bald eagles are also extremely stout birds with a body length as much as forty-two inches, a wingspan of eight feet, and weighing up to fifteen or so pounds. They are also hearty, able to live up to thirty years of age in the wild. Eagles are very swift fliers, but they prefer to soar high on thermal convection currents. They can easily reach speeds of fifty to seventy miles per hour when gliding or flapping, and can reach a dive speed of seventy-five to a hundred miles per hour. Eagles much prefer to catch fish that swim near the surface of large lakes. They can easily grab a fish with their talons at high speed and carry the fish away, with all its weight and accompanying wind resistance, at about thirty miles per hour.

Eagles are also extremely wary animals. A man on foot with a rifle usually can't get within a quarter mile of one unless he is quite skilled. You can get closer to one in a pickup than you can on foot. They will sense you, see you, and take flight. When our family couldn't scare them away with rifle shots while they were doing so much damage to the livestock, they shot them with rifles, as did Max Menzies in the picture here, which was taken in the early 1950s. He had been elected sheriff of Menard County several times by then. Realize that an eagle in your pasture could do more damage

Left to right: Ganny Wesrook, Sheriff Max Menzies and his dog Brownie, and Allan Everett

than having a pesky coyote out there. Just one eagle can easily eat a young lamb or a goat or more a day. Since bald eagles are very sensitive to any kind of human activity, most of the time they can be scared off by a rifle shot into the trunk of the tree where they might be perched. Our family tried whatever they could to get eagles to leave their land and never shot them indiscriminately. However, since they always came back to the same area where they were raised, bald eagles have been and continue to be a problem in various areas of Menard County. Back in the 1950s and 1960s, some ranchers hired pilots, took the doors off the airplanes, strapped a rifleman in the backseat, and shot the eagles from the sky after they were airborne. Even that wasn't easy, because as soon as the eagles thought they were being stalked, they would go into a speedy dive. My cousin Jim used to chase them out of his area of the country with his plane, and that worked for him most of the time.

Jim and Joan Menzies pose with their 1949 Piper PA12

It is estimated that half a million eagles called the United States home when the Pilgrims landed, but for two reasons the birds were such pests that they almost became extinct by the midtwentieth century. Ranchers shot them, but the biggest killer of the bald eagle population was DDT. This pesticide poison was not lethal to the adult birds, but it wreaked havoc with the birds' calcium metabolism by either making it sterile or unable to lay good, healthy eggs. DDT was banned in the United States in 1972. In 1918 the bald eagle was declared a protected species in the United States and could no longer be killed for commercial purposes. Then, in 1967, when there were only 417 pairs left in the lower forty-eight states, it was declared an endangered species and could not be killed for any reason, even to protect one's livestock. These protections worked so well to bring the strength of the eagle population back that in 1995 the bald eagle was removed from the list of endangered species and placed on the list of threatened species. Then, in 2007, bald eagles were further declassified to the category of least concern.

### New Technology

William was always interested in finding and using the latest technology of his day. He was ready to build new things in his workshop, where he had his own forge. Even as he got up in age, he still had a thirst for new ideas, new things, and better equipment. Even at the tender age of eighty-one, he attended the State Fair of Texas in Dallas in 1936 and again in 1937.

## The Depression

Ranching in pioneer days was tough enough, but during the Great Depression of the 1930s, it really got bad. Most of the time the Menzies kids had to do without shoes. At one point Walter had only one pair of khakis to wear, but Letha Ann was so faithful to her children that she washed, dried, and ironed Walter's khakis every night so they would be clean for him to look good at school the next day. At one point, things were so bad, if a burglar broke into their house, all he would have gotten (except for the food) was experience.

The Depression didn't just hit the big cities, but it made its scourge felt all the way out to the Texas ranch lands. It was so severe that William was unable to make his full interest payments to his brother Alex. Alex let it go, because the times were nearly impossible for everyone in the country. Rents were reduced and payments were forgiven all across the nation. Properties were foreclosed on when so many lost their jobs and couldn't pay their debts. If a person could pay something fairly consistently, that was considered a blessing. However, as soon as he could, William resumed his regular payments and continued them until Alex died. At that time Alex's wife, Mary, went to court to try to take back much of William's property because of the temporary reduction in the payments during the Depression, but the court found in William's favor. That he had resumed payments in full for so many years showed that Alex had accepted his payments and was satisfied with them over that period of time, considering the difficult circumstances. Alex had neither tried to take any of the land back during his lifetime nor had he even requested reimbursement, so the judge denied Mary's petition. This was just another difficult challenge and season of great drama for William and Letha Ann during their lifetime.

## Livestock Auctions

Before the 1880s and 1890s, when fencing came in, livestock was trail driven from southwest Texas to the railhead at Brownwood. Things fairly well stayed that way until 1911, when the railroad came to Menard. Then livestock was shipped to markets in Fort Worth, Kansas City, and points north or down to San Antonio. Until the late 1930s, livestock was carried primarily to markets in north Texas and beyond by rail, with very little competition between meatpacking companies for the price of a rancher's produce. The old concept of selling livestock at only a few large depots put the ranchers at the mercy of large meat-packing houses and forced them to bear all the expense of getting their animals to these distant places. The whole process of getting livestock to the railroad, transporting them to the major cities where the packinghouses were, unloading them there, and putting the animals up in pens, feeding them, and getting them to the sale barn was still very expensive and cumbersome. However, in the late 1930s, local livestock auction barns started springing up all over Texas. The first auction to open in the vicinity of Menard was the Abilene Horse and Mule Auction. As the name implied, it started out as a horse and mule auction but gradually widened its sales to include cattle, milk cows, goats, and sheep. It was still quite a distance from Menard, so in 1949 the Producers' Livestock Auction opened in San Angelo. It has been going strong ever since.

In the early 1950s another new marketing concept participant surfaced: the commission man. This new middleman directly bought several ranchers' livestock, hired a semitruck driver, and hauled them to the big-city auctions. With the advent of commission men, ranchers could listen to the radio market report every morning over breakfast and know how livestock was selling. When the market peaked, they would call a commission man and sell their animals from the ranch. Upon purchasing them at a discount, the commission men would whisk them off to an auction immediately, hoping for an uptick in the market. They also took the risk of a downturn and had to absorb the shrinkage of the animals en route. They always tried to sell the livestock along with what other animals they might have been able to purchase from other ranchers in the vicinity to improve their efficiency. This greatly improved things, because it meant ranchers received their check for their animals without even having to leave the homestead. However, they did have to pay the commission men. Sometimes the commission man won, and sometimes he lost. At least one commission man in Menard County was known to have caught the short end of his trade and committed suicide as a result.

The early 1950s also began another new era, that of ranchers purchasing their own medium-sized cattle trailers and taking their livestock to market themselves. With a trailer now, they were no longer at the mercy of commission men, hoping for a fair deal. They were also free of worrying about the availability of the commission men to service them on a timely basis at peak market times when everybody else wanted to move their animals. With their own trailer, all it cost ranchers to market their livestock was a day of their time, a little gas to transport the animals on the trailer behind their pickup, and the small auction-barn fee. Most importantly, whether they used their own large truck, a trailer, or the commission men, gone were the days of the long, expensive, dusty, dangerous cattle drives to distant noncompetitive markets.

In addition to forcing the meatpacking houses to come to smaller towns, the local auctions also afforded ranchers all across the state the

Antique farm machinery

opportunity to become buyers for whatever their own individual livestock needs might be. This increased the value of livestock by virtue of adding the ranchers' own purchasing power, which, taken collectively, was quite significant. The new, increased competition at the local markets, along with the decrease in the cost of transporting their livestock, had a very positive effect on the ranchers' profitability and forever changed the way livestock was marketed. All of this was very important to ranchers, because everything they did at the ranch hinged not only on their being able to sell their livestock and get the best price but also, more importantly, getting the best net after the cost of the sale. There was no sense producing animals unless they would bring a return on the ranchers' investment.

William Menzies and his sons were able to immediately take advantage of these new marketing concepts as soon as they surfaced. They already owned large stake-side trucks with which to haul their animals to market.

I remember sitting at the breakfast table with my mother's parents in Lott, Texas, and watching my grandfather, Herman Friudenberg, eat and listen intently every morning to the kind, fast-moving bass voice of the stock price announcer on the radio. Granddad would sop his biscuits in the yokes of two sunny-side-up eggs and then dip them in a mixture of clear Karo syrup and butter. He would also work in a bite or two of a red-hot link sausage and a sip or two of his black coffee. The radio announcer's kind voice was reassuring to ranchers and good to hear most of the time, as he told whether the price of goats, lambs, hogs, and cattle was up, down, or steady. These bits of news were a crucial part of the market information the ranchers gathered in order to know when the market was about to peak, so they would know when to get their livestock quickly to the cattle auction barn. Listening to the radio every weekday morning was something that happened almost without fail. I was too young to sit at the table in my great-granddad's day, but William and his friends did just about the same thing after local cattle auctions came into vogue.

In Menard, the railroad's advent in 1911 was a tremendous boon for the city and the whole surrounding area. Menard actually became the second largest livestock depot in the state for the Santa Fe Railroad. In the early 1950s, N. C. Armstrong, Clyde Dozier, and Claude Rambo opened the Menard Auction Company, which thrived until the late 1960s. It was bought out in 1968 by Charles Kothmann and Mike Murchison and did well until the railroad station closed in 1972. The company remained in operation until about 1974, when the ranchers in the region seemed to prefer a larger auction that had a railhead in San Angelo.

## The Great Screwworm Plague

Without a doubt, few things changed the ranchers' lives for the better more than gooseneck trailers, pickup trucks, and local cattle auction barns. However, the eradication of the New World screwworm fly certainly trumps all else in the area of improving the quality of life and work for the ranchers. Amazingly, most of the crucial scientific discoveries related to the worm's demise came from research conducted in Menard County.

For the better part of William's ranching life, he and his boys, like all the other ranchers, had to ride the range daily during the summer months to search out infested livestock they called "wormies." It was a monumental task to inspect and doctor their animals for the parasite known as a screwworm. Ranchers spent so much time in the saddle back then that they could easily wear out the rowels of a good set of spurs over a lifetime. My great-uncle Tom Jacoby did exactly that. The plain truth is that this was as much a part of a ranchman's lifestyle as breeding, feeding, branding, shearing, and marketing livestock.

A female screwworm fly lays her eggs in an animal wound.

The worms were actually the larvae (or maggots) of this fly species, and they fed on live animal tissue. Screwworms were transmitted to animals when a female fly laid 250 to 500 eggs on any kind of superficial wound the animal might have. The larvae hatched and immediately burrowed into the animal's tissue. Many wounds were susceptible to this infestation, such as fence wire and shearing cuts, tick bites, tail bobbing, dehorning, branding wounds, etc. The fly's primary target, however, was the wet naval umbilical cords of newborns. The first infestation and its odor attracted other screwworm flies to initiate multiple infestations. Screwworm flies were quite stout and could easily travel twelve miles.

The effect of an infestation on animals was difficult to detect immediately, but the stench was soon a giveaway. The larvae fed inside the wound, gradually widening and deepening it. Soon a blood-tinged, foul-smelling fluid would begin to seep out. As many as a couple hundred larvae would pack themselves inside a wound. Infested animals became lethargic and trailed off from the herd to find a shade tree to lie under. A declining appetite, much discomfort, and a decrease in milk production were common symptoms with this kind of worm infestation. Animals that went undetected and untreated died in seven to ten days from toxicity and further infestations. All warm-blooded animals—sheep, goats, cattle, rabbits, deer, birds, and even human beings—were affected by these parasites.

This fly's larva was so virulent that they nearly wiped out the deer population in the 1930s and 1940s. Fawns were born mostly in May and June, which happens to be the height of the fly season. Deer move so swiftly and wildly that there was no way for the ranchers to treat them for these deadly parasites.

It was the cowboy's job to first identify wormy animals, usually by the smell. He would then rope and tie them down, or just physically hold down a smaller sheep or goat, while he cleaned out the wound and treated it. This was momentarily painful for the animals, but these efforts were definitely lifesaving in the long run. For the most part, ranchers used something as simple as a loop of baling wire to rake out the maggots and then apply the medicine. Chloroform was used at first, and then two or three other medicines were developed by the US Department of Agriculture laboratory in Menard. These medications or preparations were terribly smelly creams that both treated and protected the wound. Keep in mind that these treatments had to have a scent powerful enough to repel the other flies, and that called for a mighty strong odor.

The Menzies ranches worked closely with USDA scientists by trying all the newest techniques, medicines, trials, and applications. Doctoring wormies was, no doubt, the pits of the ranching business at the time. A cowboy with a weak stomach usually lost his lunch while doctoring screwworm wounds. Even if he got past doctoring the wound, he had a devil of a time trying to get the smell off his own body. My cousin Jim Menzies did his share of doctoring wormies when he was growing up, and he said he would "come home at the end of a day of treating animals and try my best to wash up before supper, but you just couldn't get the smell off your hands."

The next day they had to go out all over again, find more wormies, and treat them. Later on, they started bringing the wormies back to the ranch barn and putting them in a "wormie trap" or small corral so they could treat and daily monitor the animals more easily and effectively. All newborn livestock especially had to be checked for eggs in their navels. It was amazing how a little fly with only a twenty-one-day life cycle could wreak so much havoc.

In the early 1900s screwworm flies began attacking herds in the Southwest. None of the techniques used to control these parasites was successful, and it got so bad that it, along with the added difficulties brought on by the Great Depression, made livestock production almost

One of the huge fly traps required by the USDA in the
1930 screwworm fly eradication program

unprofitable. This fly single-handedly decimated livestock herds all across the entire South, including Texas, Mexico, and Central and South America.

Ranchers first experienced increasing trouble in 1912 and requested assistance from the Department of Agriculture in 1913. At first, two entomologists and a biochemist came to Menard, studied the problem, and made some recommendations. Their solution was to put flytraps all across the county and have the ranchers limit earmarking, castration, dehorning, etc. during the seasons of the screwworm's greatest activity. These directives were followed all over Texas, especially in Menard, where the research was taking place. A total of 664 enormous flytraps were constructed and placed strategically across Menard County's 156,000 acres. The experts also recommended putting tar oil on animal wounds to repel the flies, as well as recommending benzal be used to kill the maggots.

In 1929, the Bureau of Entomology and Plant Quarantine established a research laboratory in Menard to assist in the implementation of these directives. This program only lasted from 1929 to 1933, when sufficient statistics were accumulated to show this to be an ineffective solution to the

Scientist H. E. Parish stands beside 33,508 quarts of screwworm flies caught during a ten-month period in 1932 by ranchers in the 664 traps spread across Menard County.

A Progressive Rancher

screwworm problem. Research became continuous at the Menard lab, and four entomologists—R. C. Bushland, E. F. Knipling, R. Melvin, and H. E. Parish—were assigned to find a solution to the screwworm problem as well as to formulate a better wound protectant.

In 1937, the Screwworm Research Lab found a new home, dubbed the "Stink House," which the USDA built on the old Fort McKavett Road about three miles west of Menard, on land originally owned by Sod Crowell. Later, the majority of the land was purchased in 1951 by Alex Menzies, William's son. Shortly thereafter, Alex's son Jim and his family began living there and are still on the ranch today. Jim himself developed a keen interest in agricultural science, and upon graduation from Texas Tech University, he worked for ten years as a scientist at the Texas A&M Experiment Station near Sonora. His brother, Dr. Carl Menzies, also had a strong desire to improve agriculture, and upon graduation from Texas Tech and other schools, he had a teaching career at various colleges. In 1971, he was named director of research for the Texas A&M University Research and Extension Service in San Angelo, where he was in charge of the agricultural research programs and experiment stations in San Angelo, Barnhart, and Sonora for twenty-five years.

The original Stink House still stands today, preserved as a Texas historical site, but relocated to the USDA scientific research site in Kerrville.

Delaine ram skull

This little building—the Stink House—that was originally built by the USDA for screwworm research was only sixteen-by-thirty feet and had a longitudinal partition down the middle of the structure, with a screened vestibule on the front that allowed entry to either of two backrooms. The house was sturdy and simple in design. One of the backrooms was filled with number-three washtubs containing an odorous slurry of dead meat, worms, and flies. It did, of course, what it was designed to do: propagate huge numbers of flies. The other room was dedicated to testing and manufacturing insecticides as improved worm killers.

No one had to tell these scientists why the building was called the Stink House. They received the full benefit of those pungent aromas every hour of every day. There was no central heating or air-conditioning system, so these researchers paid a real price every day they went to work. Few knew or appreciated the dedication of these scientists who worked long hours every day in terrible conditions, doggedly determined to find cures for this plague on society.

In any lab environment (and especially one having open slurries of rotting animal flesh, flies, and maggots), there were many dangers from germs and diseases in the vicinity, like anthrax and cholera, for example. Those threats never deterred them. They all knew that the anthrax germs could lie dormant in the ground for twenty to twenty-five years and, without any warning, become active again, quietly killing livestock and people. Then the germs would return to the soil, encapsulate themselves, form a spore, and go dormant again. The only way to known if any livestock have the live virus is when the animals start dying.

## Other Deadly Diseases

Cholera is the other deadly disease that can appear out of nowhere, just like anthrax. Every scientist entering into this kind of work accepted, like a soldier, the real possibility that he might lose his life in service to others while researching various cures. These USDA researchers deserve far more credit than they ever received, I assure you.

A case in point was my uncle George Craigmile "G. C." Menzies Jr., grandson of William and son of William's first son, George. He first served his country as a naval officer during World War II, working with malaria and epidemic control units in the South Pacific. Upon his return from the war, having already earned his bachelor's and master's degrees in entomology from Texas A&M University, he went to work as an entomologist for the Texas State Department of Health in Austin. In the mid-1950s G. C. volunteered to research rabies on domestic bats in the caves of south Texas. His field of research included collecting bats to send to the Austin lab and tagging them to discern migration patterns. As far as anyone knows, he was never bitten or even scratched by a bat. However, he did have a war-related lesion on the back of his neck, and his colleagues believed that the rabies virus was possibly transmitted in an airborne manner either through this lesion, his eyes, or his nose while he was in one of the caves. They confirmed conclusively that G. C. contracted rabies, which was the cause of his untimely death on January 4, 1956. He died at the age of forty-one, a great loss to his wife, his four children, and his extended family. G. C. lost his life doing research on a disease for the benefit of his fellowman.

History is replete with examples of scientific triumphs that have benefited society so wonderfully. But these benefits often came at the cost of the lives of those who diligently sought and found them. Yellow fever was a feared disease in the tropical and subtropical regions, capable even of invading and devastating the temperate zones. For over two hundred years, it was considered one of the greatest plagues in the world. A New Orleans scientist named McNeese died while investigating what made yellow fever such a virulent killer. Deadly epidemics had stopped the wheels of industry and trade in

A plaque at the State Health Department in Austin testifies to G. C.'s dedication

England, France, Italy, Spain, and the United States in the late 1800s. As late as 1905, there were yellow fever epidemics in the United States in twenty of the largest Northeastern cities and some Southern cities.

During the Spanish-American War, Maj. Walter Reed was appointed to head a research committee charged with finding the cause of typhoid fever. An outbreak of typhoid among US troops in Cuba had been devastating. To complete his research he conducted tests with human volunteers, many of whom were his coworkers. In order to confirm his research, many sacrificed their lives. However, a cure was found. Reed discovered that the virus infects monkeys, rodents, and opossums but was only passed on to humans by mosquitoes. Thanks to him and others who conducted research on his findings, typhoid fever and malaria were made preventable diseases. As a result, construction of the Panama Canal became possible. The huge military hospital in Washington, D.C. was named after him.

We can only thank these scientists and God for the discoveries that have benefited mankind over the years. These scientists and others like them today continue to make these remedies possible often at the risk of their own lives and health. Only a few are mentioned here.

## Defeating the Screwworm Fly

It is important to remember that it was in Menard, Texas, in 1941 that these scientists developed the much-improved remedy for screwworm infestations known as Smear 62—a smelly black cream that both treated and protected an animal's wound. It could better kill the deep-boring worms and also gave protection against reinfection. Most importantly, it was in Menard that crucial information was discovered along with the formulation of the ultimate plan for defeating this parasite.

At the center of solving the screwworm fly plague was one particular scientist: Dr. Ed Knipling. He grew up on his father's farm near Port Lavaca, Texas, and saw firsthand the destruction caused by insects, especially screwworms on his father's livestock as well as the boll weevils on their cotton. This kindled an intrinsic desire within him to study agriculture and entomology at Texas A&M. It was there he discovered the tremendous effect insects had—good as well as bad—on plants, livestock, and humans.

Knipling's first job with the USDA was trapping and monitoring the screwworm fly population in Menard. In 1935 he began working with

Dr. Raymond Bushland, doing this research as well as finding new treatments for livestock that had already been infected. It was here that Knipling realized that screwworms could never be controlled this way. Over the next several years, it occurred to him that somehow they needed to stop the flies from laying eggs rather than treating the wounds. The sterile fly technique he discovered came about from several observations he made in the 1930s.

He noted that the male files aggressively mated with several females, but the female flies would mate just once. He also observed that the screwworm fly population was very low on a unit-per-acre basis. Simultaneously and independent of Knipling's observations, Bushland and others developed a method to artificially mass produce large quantities of flies in the lab for research studies. This mass production of flies and knowledge of how the flies mated gave impetus to the idea of raising genetically defective flies in the lab and releasing them en masse to mate with the female flies in the field. The scientists knew that the resultant eggs would never hatch. They theorized that this practice would diminish the population of each subsequent generation finally to extinction. The only problem was that Knipling did not know how to induce a genetic defect into the male flies. He worked out his idea of population control with mathematical models predicated on the laws of probability. Then his idea sat dormant for ten years, interrupted by World War II, when he served in the army and researched mosquitoes and similar insects to help in the war.

In 1946, several small labs were relocated at a consolidated experiment station in Kerrville, Texas, where twelve or so entomologists continued the screwworm research. The Stink House lab in Menard was moved there as well. Later, in 1950 the researchers developed the formulation called EQ 335, which was a vast improvement over Smear 62. They then developed a jellylike substance known as 1062, which had even better qualities. All of this certainly helped, but it did not eliminate the problem.

At about this time, Knipling was headquartered in Washington, D.C. as chief of the Insects Affecting Man and Animals Research Branch. He had not given up on his idea of releasing sterile male flies as a way to annihilate the screwworm fly.

During the war, atomic radiation was discovered, and Dr. H. J. Muller at Indiana University discovered how to sterilize fruit flies by exposing them to X-rays. Knipling wrote to him and shared his idea of sterilizing screwworm flies and received a very enthusiastic response.

Drs. Edward F. "Ed" Knipling (left) and Raymond Bushland (right)

The concept was put to a large-scale test by Knipling's former Menard colleague, Dr. Bushland, who was at the Kerrville research facility. The rest is history.

Knipling and Bushland felt sure that screwworm flies could be eradicated by introducing enough sterile males to outnumber the fertile males, thus breaking the fly's reproductive cycle. This would end the use of livestock and wild animals as hosts for its deadly parasitic larvae. In the Kerrville Lab cages they conclusively proved that the sterile males were equally competitive with their fertile counterparts in the wild.

The next step was to prove that this system of eradication would work in the field, and a controlled test was conducted on Curacao, an island about fifty miles from Venezuela. With proof in hand, millions of male flies were harvested, radiated, and air-dropped all over Texas. But there was not enough government money to cover the enormous cost of this part of the eradication program, so it fell to the local ranchers to raise the millions of dollars necessary to complete this treatment.

Just think of the job they had to convince the ranchers in the saddle riding the range that the screwworm problem could be solved by sterilizing flies with X-rays and air-dropping them out of planes in small cages by parachutes. They were recommending that they be dumped out by the millions all over the state. That was really a stretch. In any event, the ranchers grasped the picture and sponsored fund-raising drives all across the state. Local leaders including several of my uncles went from ranch to ranch, door to door in the residential areas, and from business to business in the cities and got the job done. This great financial need was laid at the doorstep of these ranchers, mind you, in the midst of a severe drought in the 1950s. It was probably the most difficult time in Texas ranching history. Regardless of the circumstances, the ranchers sacrificed and donated thousands of dollars to make up the millions that were needed.

A huge debt of gratitude goes to Jerry Puckett, a rancher from Fort Stockton, Texas. Puckett served in World War II, graduated from Texas A&M, and then, in addition to ranching a huge spread, served on the board of the Southwest Animal Health Research Foundation (SWAHRF) for many years. He was personally responsible for raising over four and a half million dollars for this great cause. He said, "A lot of 'em told me you can't do it and don't even try. I never really thought I could do it either, but there was something down in me that said I ought to at least try. I remember riding down two horses a day back then (doctoring wormy animals), one in the morning and one in the afternoon. My dad even had a dog that would lay on its back in the bed of the pickup, and it would raise up and bark when the truck got anywhere near a wormy animal. It was a fight, but I got a lot of help."

The organizers suggested that the ranchers donate ten cents per head of sheep and goats and fifty cents per head of cattle. These donations were strictly voluntary, but most everyone did their part. They just anted up. However, after all they did, there still wasn't enough money, so they had to go back to the ranchers and ask them to dig deeper despite the terribly difficult economic climate. But that didn't do it either, so they went back to the ranchers a third time. Thank God for Jerry Puckett and those

Jerry Puckett

west Texas ranchers who never gave up until they got the job done! They had that same never-give-up Texas spirit about them. It took faith, determination, and an acceptance of individual responsibility.

Jerry is eighty-four years old today, still married to his bride of sixty-four years, and still going strong, running the day-to-day operations of a large ranch in Fort Stockton, Texas.

SWAHRF worked tirelessly on the screwworm eradication project into the 1960s and 1970s, and it also secured federal and state funding for the venture. The Texas sheep industry was the first to donate to this much-needed program. More than eight million dollars was raised by SWAHRF from individual ranchers and the state and federal governments. To this day, the screwworm eradication project remains the most effective program ever conducted to benefit American agriculture and especially livestock production. As much as the screwworm eradication project helped domestic animals, as a side benefit, it also helped the wildlife. In fact, the white-tailed deer population exploded immediately after the demise of the screwworm fly.

With the money these ranchers raised, millions of male screwworm flies were harvested, irradiated, and air-dropped. By 1982 the screwworm fly was eradicated from the southwestern United States. The success of this program also helped ranchers to learn the value and importance of agricultural research.

After the initial success of the program, virtually every Texas rancher cooperated with the USDA by carrying special kits in their trucks to collect the larvae from any further animal infestations. They turned these in at local labs for testing to ensure that the maggots they had found were not from screwworm flies but from a less dangerous species. If a new screwworm outbreak was discovered, aerial sterile fly drops were ordered until the infestation was eradicated. These efforts were coordinated with Mexico, and screwworm flies have now ceased to be a factor north of the Isthmus of Panama.

Thanks to these scientists and the financial contributions and untiring cooperative efforts of thousands of Texas ranchers for seventy years, this program was so effective that few people today even know that there ever was such a thing as the pesky, deadly screwworm fly.

The screwworm fly's demise, however, detrimentally affected the Texas horse market for a time because fewer people rode their ranges to check for screwworms, and thus fewer mounts were needed. On the other hand, with these parasites banished from the territory, productivity increased on the ranches because fewer cowboys were needed to ride the range to battle this problem. While ranchers were searching and doctoring wormies, they couldn't be building barns, improving their corrals, branding animals, or building fences.

One other benefit of this program was the discovery of the better wound treatments, all of which occurred in that little building in Menard known as the Stink House. It was there that scientists discovered the most beneficial and effective program to help American livestock production, agriculture, and wildlife. The secretary of agriculture, Orville Freeman, described this research program as the "greatest entomological achievement of the century." For their crucial discoveries and tireless efforts, Drs. Edward F. Knipling and Raymond C. Bushland were joint recipients of the World Food Prize in 1992. Knipling, for this and many other achievements too numerous to name, also received the National Medal of Science, the President's Award for Distinguished Federal Civilian Service, and the USDA Distinguished Service Award.

## Droughts

Another problem that has plagued Texas over the centuries is droughts. Texas has never had an overabundance of rain, but droughts are chronic, prolonged periods where there is a lack of rainfall or extremely small amounts. A terrible drought, lasting several years began in 1919 and another started in 1950 and lasted seven years. During the latter, many West Texas families sold out and left. Recurring droughts have adversely affected the agrarian economy as much as the screwworm plague. But even with all the problems Texas has experienced over the years, it still produces more beef than any other state in the union.

## The Boll Weevil Blight

For most of the nineteenth century and the first quarter of the twentieth, cotton was ubiquitous in the South. As a cash crop, it was king. However, in about 1890, unbeknownst to American cotton producers, a hoard of the most destructive pests ever to attack the US cotton industry began marching their way. By 1894 the boll weevil beetle had migrated from Central America, completely across Mexico, to Texas, specifically to the Brownsville area.

This little beetle is only about a quarter of an inch long (including its snout, which accounts for a third of its length). It lives almost entirely off of the flowers and buds of cotton, the only known host to this beetle. Since the insect had no known natural enemies or program to defeat it, once it was here, hoards of weevils grew exponentially. They were capable of traveling from forty to sixty miles a year, and had infested the entire US cotton-producing region by the 1920s. They truly wreaked havoc on the agricultural industry. No one knew how difficult it would be to find a solution to the problem or how long the research was going to take. Historians and entomologists place the boll weevil second only to the Civil War as the most significant cause of detrimental change impacting the Southern states. Over the last century, the boll weevil has caused production losses and created control costs estimated at better than $22 billion. In recent years the boll weevil has cost the South in the neighborhood of $300 million per year.

In the 1920s cotton was a popular crop in Menard County. Five cotton gins were built to handle all the cotton: one near Fort McKavett, one in town, one each at the five- and ten-mile crossings of the San Saba River, and another at Hext. Thus they completely spanned the county. The one in town was owned by Mayor Pete Andregg, who incidentally and tragically lost an arm while operating the gin. What was called "dry land cotton" had been planted not only in the valleys of Menard but also on many hillsides as well.

In addition to the problems the boll weevil and the Great Depression presented, Southern farmers had another massive problem to contend with. Simultaneously, all the Southern states had been growing more and more cotton until the mar-

Boll weevil

Cotton cart

ket was flooded. In 1931, just as Texas alone produced better than five million bales a year, every cotton farmer in the country began to lose money. Market prices suddenly dropped like a brick to just six cents a pound, which was several cents less than the production costs. At this point, Texas hosted a cotton conference in Austin, where the representatives called for each state to restrict production. That agreement limited Texas farmers to planting cotton on no more than 30 percent of their property. All the other states passed similar laws but worded them to take effect only when and if 75 percent of the cotton-producing states actually joined in the effort. Of course, this meant that Texas was the lone state where this law took hold immediately. Later on, the New Deal included an Agriculture Administration Act that limited all the states alike, but Texas farmers suffered severely in the meantime.

With the combination of an increased lack of rainfall, a flooded cotton market, the Dust Bowl years, production restrictions, the loss of valuable top soil over the decades, and the Great Depression, cotton production and the cotton gins completely vanished from Menard County. The county's farmers turned to more productive crops.

Historically, the boll weevil became a significant problem for cotton farmers all across the South in the 1920s, but when the Great Depression hit in the 1930s, the pest really exacerbated the situation. Different treatments were tried, but it wasn't until after World War II that pesticides like DDT were developed that enabled farmers to once again grow cotton profitably. However, even this came with a hefty price tag and was done at great risk to the environment.

In the 1950s two events contributed greatly to the obliteration of the boll weevil. First, the cotton industry was impacted with such tremendous economic challenges due to droughts and boll weevil infestations that it became evident that to reduce losses and costs, elimination of the weevil had to become a high priority. Simultaneously, a major research revolution to eliminate certain insects was started by Edward F. Knipling, a pioneer in the field because of his efforts with screwworm fly program. His success with that program implied that a similarly successful program could be conducted against the boll weevil. The Southwest Animal Health Foundation that had facilitated eradication of screwworm flies served as the template for an organization that would do away with the boll weevil.

In 1961, the Agricultural Research Service (ARS) built the Boll Weevil Research Laboratory in Starkville, Mississippi. There, researcher Dick Hardee made some important findings between 1965 and 1969. He developed the material with which to test various compounds on boll weevils. A few years later, while working as a graduate student for ARS at Mississippi State University, Jim Tumlinson discovered a successful bait, a pheromone lure, that combined four compounds.

Another significant improvement contributed by the various ARS labs was the design and production of an inexpensive and far more effective detection trap. Since 1987 over twenty million of

A boll weevil trap

these traps have been manufactured and used in the South. The pheromone in these traps attracts the weevils. Farmers can then calculate their numbers and determine the level of infestation in a particular field or area. Once infestation reaches a certain level, it targets and triggers the use of insecticides to eliminate the pests.

Heavy applications of various expensive insecticides were tried, including DDT, until a much better poison was found. In the 1990s it was discovered that much lower dosages of Malathion were more effective, less expensive, and much less dangerous to the environment. Malathion was actually found to be one of the safest insecticides available. Furthermore, it is excellent for use in areas that are also occupied by people, birds, and domestic animals. Because of Malathion's potency, ultrasmall volumes of the pesticide are needed. Just ten to sixteen ounces of it (about the amount of liquid in a soft-drink can) can treat an acre of cotton.

Researchers, including lead entomologist Dale Spurgeon with the ARS Research Unit in College Station, Texas, believe that there are only a few major issues to research in order to completely wipe out the boll weevil from the United States.

Farmers, agronomists, agricultural economists, and extension entomologists in cooperation with USDA scientists and agricultural experiment stations worked together over many years to plan, create, and implement the Texas Boll Weevil Eradication Program. Just as in the case of screwworm eradication, it took a huge, cooperative effort between government scientists and local farmers over a protracted period of time to bring about the possibility of the complete elimination of the boll weevil. This potential success is a tribute to the sacrifices, faith, hard work, and cooperative spirit of everyone involved.

Most of the knowledge about the boll weevil's basic ecology and biology was discovered and developed in Texas. The delayed planting technique for the rolling plains, realizing the need for early destruction of cotton stalks after harvesting, the trap index system for weevil infestations, the release of sterile males to reduce the insect population, the development of short-season-determinate cotton varieties, and the suppression of the beetle through the use of insecticides in the fall were all researched and developed in Texas. In addition, most of the development and plan for the use of the boll weevil trap and its tactical use as a monitoring tool was likewise researched and developed in Texas. All these practices and devices were the primary components that were later integrated into today's successful pest management programs. After 1978 the Boll Weevil Eradication Program made it possible for cotton to resume full-scale cultivation.

In 1993 the Texas legislature established the Texas Boll Weevil Eradication Foundation (TBWEF). This nonprofit organization is managed by cotton producers and oversees the implementation of the boll weevil eradication program in Texas. This program first began in 1978 and has been very successful. In 2009 the TBWEF announced that this destructive insect's impact had been reduced to the extent that cotton farmers in Texas were able to set all-time-high production records in three of the most recent six years. Texas still produces more cotton than any other state. To date, the Boll Weevil Eradication Program is also being conducted successfully in seventeen Southern states.

Consider for a moment the impact of this program on the state of Georgia. Prior to the beetle invading the state in 1915, the state's annual cotton production was 2.8 million bales. A mere ten years later, production fell by 80 percent to 600,000 bales and continued to plummet to a low of 112,000 bales in 1983. The eradication program began there in 1987, and ten years later production rebounded to 1.66 million bales. The cotton industry's farms, gins, warehouses, cottonseed mills, and textile mills also generated 53,000 jobs and contributed a positive economic impact for the state of $3 billion a year.

The boll weevil is not yet completely eradicated, but hundreds of millions of dollars are being spent annually to hold its population at bay. Although full cotton production has been facilitated and much about the boll weevil has been learned, research continues in order to completely eliminate this destructive insect from the United States just as we have the screwworm fly.

Chapter 9

# A Blessed Union

THE COMMAND FROM ON high was, "Be fruitful and multiply; fill the earth and subdue it; have dominion" (Genesis 1:28). All this William and Letha Ann Menzies did. In addition to subduing the earth and the beasts that roamed it, their union was also mightily blessed with eight wonderful, able children. All of them were successful in business and family life. Every one of their children were also blessed with character and a good work ethic. Not one ever darkened the door of a prison for

Left to right: William Menzies Sr., Max, Alex, Letha Ann, George, William Jr., Letha, Agnes, Pearl, and Uncle Alex. Walter was born after this picture was taken in 1907.

The Menzies boys (left to right): Alex, Bill, Max, and Walter. George not in this picture.

wrongdoing. The firstborn was George Craigmile Menzies (my granddad). He married Ella Bertha Pfiefer on June 11, 1913. They ranched in

George Menzies

Ella Menzies

Menard County, first by helping William at the headquarters ranch and then out at the Gap ranch.

George and Ella had two sons: George Craigmile "G. C." Menzies Jr. and Perry Phillip Menzies (my dad). Things were going only too well for

G. C. Menzies

Perry Menzies

them until a couple years after Perry was born. It was then that his father, George, contracted a very serious and mysterious disease that the doctors could not diagnose, so they certainly knew of no cure. From the symptoms he was having, his disease was only much later thought to have been sugar diabetes. Several other members of the family have had it subsequently, including William and my dad, Perry. However, it had little effect on them as a result of today's much-improved medical techniques. In 1919, even if they could have diagnosed it at the time, they knew nothing about any remedies. Doctors now know that diabetes causes strokes, nerve pain, kidney failure, blindness, and peripheral vascular disease, which keeps the blood from reaching the extremities, the hands and feet. Today, when the limbs turn black from the lack of circulation, the affected limb is amputated before gangrene can set in. This is known as end-organ damage, where the main organs of the body begin to shut down, causing horrific pain and physical problems.

At that time, the family was sixteen miles from town at the Gap ranch with no telephone and no transportation other than a wagon. It is not known if George was capable of harnessing a team of horses and driving back to town at the time. In the final stages of the disease's devastation on him, in 1920, he took his own life. George Jr. was about five years old and Perry was three at the time he died. The passing of the firstborn son hit everyone very hard and was so painful for the entire family that it was never much discussed.

Agnes Craigmile Menzies, the second child of William and Letha Ann, was born on July 25, 1891. She married William Raymond Walston on

Raymond and Agnes Walston

Raymond Roy and Willie Lee Walston

October 29, 1919, after he returned from World War I. They ranched south of Menard off Highway 83. Agnes was a rural schoolteacher for ten years, and Raymond served as a trustee on the Palmer school board in addition to ranching and operating a general store. They had two children, Willie Lee and Raymond Roy.

Left to right: Raymond Roy Walston, Roy Jacoby, Willie Lee Walston, and Ray Jacoby at Granddad's house.

William "Bill" Menzies Jr. and Letha Ann Menzies were the twins of the family, born on August 26, 1894. Bill served in World War I and later

Bill Menzies

Anne Menzies

Billie Ann with her lariat and some kid goats at the Gap ranch in 1929

was the first in the family to graduate from college (Texas A&M University). Bill and his future wife, Anne Crawford, attended Kitchens School growing up. Anne's family owned a couple sections of land on the north side of the San Saba River contiguous to the Menzies homeplace. Bill and Ann were married on March 15, 1926. After their marriage, they ran the Gap ranch in Menard for a few years, then they ranched in Kimble County and finally in the Big Bend area before returning to the

Billie Anne, Iva Marie, and their dogs

headquarters ranch years later, when the estate was settled. They had two daughters, Billie Anne and Iva Marie.

Letha Ann Menzies, Bill's twin sister, married Thomas J. "Tom" Jacoby on June 30, 1926. She was a rural schoolteacher for a good many years before marrying. They ranched in Menard, Schleicher, and Kimble counties and had three sons: twins Ray and Roy and then Phillip.

Tom and Letha Jacoby with the twins

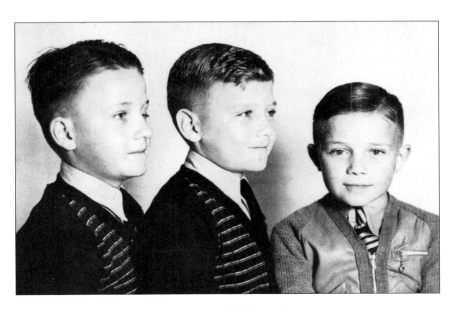

Ray, Roy, and Phillip Jacoby

Roy Jacoby, Billie Anne Menzies, and Ray Jacoby

William and Letha Ann's fourth child, Pearl Menzies, was born August 17, 1896. She married Irby McWilliams on November 25, 1920,

Irby and Pearl McWilliams

# A Blessed Union

Mary Pearl McWilliams, Duery Menzies, and M. D. McWilliams

and they ranched in southeastern Menard County. Irby was a director of the Hill Country Hereford Association. They had three children, A. J. (Alexander John), Mary Pearl, and M. D. (Menzies Davis, after Pearl's maiden name and Dr. A. E. Davis, who delivered him).

Jack and A. J.

Left to right: Vernon Crawford, A. J. and M. D. McWilliams, and William Menzies

The fifth child, Alexander "Alex" Littleton Menzies, was born September 18, 1899. He attended John Tarleton State College (now Tarleton

Alex Menzies and a soldier practice close order drill

# A Blessed Union

Alex and Max Menzies

State University), where he played football with his brother Max and was a member of the corps of cadets. Tarleton was a military junior college at the time, and Alex achieved the rank of lieutenant colonel and commander of the corps, the first one they ever had at Tarleton. Alex married Marguerite Watson on November 27, 1925, and they had three sons: Alexander Littleton Jr. (known as Sonny), Carl Stephen, and James "Jim" William Menzies.

Alex and Marguerite's family first ranched at Jackson Hallow in Menard. Then they ranched in Stephenville until 1939, when they came back and ranched the Gap in Menard. In 1945 they bought the Elm Creek ranch. When the estate was settled, they received a ranch at the Gap, and in 1950 bought the ranch on Four Mile Road.

Alex and Marguerite Menzies

Left to right: Jim, Carl, Alex, and Sonny get ready to swim in the water tank at the Gap ranch

Left to right: Aunt Frances and Carl and Alex Menzies

# A Blessed Union

Maxwell "Max" Duery Menzies was born June 14, 1902. He played football for Menard High School in 1922, the year they became state champions. He later graduated from Tarleton State College, where he lettered in several sports and was captain of the football team. A good many

Max

Kittie Sue

years later he was elected to the Tarleton Athletic Hall of Fame for lettering in four sports: football, basketball, boxing, and track. Max also ranched in Menard County, where he was elected sheriff three times from 1941 to 1953. He married Kittie Sue Harrison on August 27, 1933, and they had three sons: Max Duery, William Harrison (known as Duck), and John Marion.

At left, left to right: Bethel, Underwood, and Max

Max in the sheriff's office

Duck, Duery, Max, and John Menzies

Walter and Hazel Menzies     Scotty and his best buck ram

Walter Menzies, the youngest son, was born May 14, 1910. He married Hazel Whitley and they also ranched in Menard. They acquired several tracts of land out at the Gap and bought another ranch off Highway

Scotty and Walter at the Gap ranch

83 as well. At the time of this writing, Hazel has celebrated her 102nd birthday. She is still able to see, talk, walk with a walker, and clearly recall the days of old. They had one son, Walter Scott "Scotty" Menzies Jr., who attended Texas A&M.

## Ranching in Menard

William and Letha Ann's eight children became the ranch, farm, and home-industry hands needed to run the family's operations while they were growing up. All the kids, if they didn't learn anything else, certainly learned what a day's work looked like. William let everyone work just half days back then—whatever twelve hours they wanted. However, those days usually started well before daybreak and ended after nightfall.

### Sandlot Football

Another little story about the Menzies boys was recounted to me by Scotty Menzies. When he was in his twenties, he was hanging around town one day and encountered an older fellow from London, Texas. He introduced himself of course as Scotty Menzies, and the man said, "I know some Menzies." The man recalled that London once played against Menard in a football game on a Sunday afternoon, and London was fairly well beating the tar out of Menard. Then Alex and Max Menzies, who were good-sized men, rode up on their horses, jumped down, took off their spurs, and joined in the game. "From that moment on," the man said, "London took the drubbing."

Today, playing tackle football might not sound like much relaxation on a Sunday afternoon. But for the Menzies boys, it was probably a lot easier than what they had been doing all week on their ranches.

### Supporting the Children

William and Letha Ann supported their children by helping to further their educations. They also encouraged all their grandchildren and great-grandchildren in their 4-H pursuits. When his children, grandchildren, or great-grandchildren won various awards with their animals at the county fair, William would always take the time to congratulate them personally and give them a five-dollar bill (which means that he gave his children about $116 in today's dollars, according to MeasuringWorth.com).

The 1948 Texas State Champion Judging Team: Jakie Landers, Carl Menzies, W. H. Lehmberg (coach), Fred Sutton, and M. D. McWilliams. This photograph was taken just before they left for the national competition in Ohio. Another of William's grandsons, Raymond Walston, was on the Texas Champion Judging Team for Menard in 1941.

Christmas at the Gap ranch in 1933. Back row: Letha Ann and William Menzies. Middle row: Raymond Roy and Willie Lee Walston, Mary Pearl and A. J. McWilliams. Front row: Phillip, Roy, and Ray Jacoby and Billie Anne Menzies.

Walter Menzies in 4-H with his prize-winning Hereford in 1930

M. D. McWilliams with his prize-winning Hereford

M. D. McWilliams with his Hereford and ribbons

Duery Menzies with his prize-winning sheep and Bud Nolan, the county agent. After college, Duery served as a county agent in Fredericksburg for twenty-four years.

John and Duery Menzies with a lamb

He also made every effort to attend local football games and dearly loved to watch his kids and grandkids play the sport.

Sometimes the Scottish stereotype brings the connotation of being tight with one's possessions. Not so with William and Letha Ann. These were very generous, good people. When each child married, the parents gave them a Jersey cow and a calf, a coup of chickens, a hog, a mattress, down pillows, and many linens. They took the children in whenever they needed a roof over their heads, gave them a job if they needed one, gave or loaned

Menzies clan picture around 1945. Back row: Tom Jacoby, Agnes Walston, Letha Jacoby, Ann Menzies, Pearl McWilliams, Perry and Mary Louise Menzies, Ella Menzies Kelly, Marguerite Menzies, Winston Menzies, Raymond Walston, and William Menzies. Middle row: A. L. Menzies, Ray Jacoby, Steve Menzies, Roy Jacoby, Bill Menzies, Jim Menzies, and Walter Menzies. Front row: Marie Scruggs, Carl Menzies, M. D. McWilliams, Duery Menzies, and Scotty Menzies.

them money, and would let them use anything on the place they might have needed from time to time. If they had some bad luck, they would send them a bull, a boar, a turkey, or sometimes all three. They visited them when they were sick, and of course they loved their grandbabies best of all.

## The Golden Wedding Anniversary

In 1938 William and Letha Ann celebrated their golden wedding anniversary. It was quite a gala affair with an evening dinner in the Crystal Ballroom of the Bevans Hotel in Menard (back then there were chandeliers in the ballroom). The family hosted over a hundred dinner guests. Mrs. Noyes came all the way from Florida. Dr. and Mrs. McKnight were also there, along with a great many kissing cousins and a lot of friends from the community. William was very proud that night. His bride wore a

William with his first grandson, G. C.

black lace dress with jeweled combs in her hair. She also wore gold earrings and a corsage. It was truly a splendid evening of remembrances, fellowship, and fine food for everyone.

The keynote speaker for the evening was one of the county's leading citizens: William's good friend and attorney, Fred T. Neel. At the time, Neel was serving as mayor, a school board trustee, an elder of his church, and a leader in several other civic and business associations. That night he related many happy, difficult, and little-known stories about the early struggles of William and Letha Ann's pioneer life. What everyone realized that night was that neither all the gold on earth nor the value of all the songs sung in heaven could have purchased the joys these two shared working side by side, rearing their family and building their ranches. The couple was truly celebrated by all their friends and family. Neel's remarks were also published in the *Menard News*.

Chapter 10

# The Latter Days

**W**ILLIAM STARTED RANCHING IN Menard with little more than a loan on a piece of land, his tool chest, a herd of horses and mules he had built up by trading, a healthy load of determination, and his faith in God. When he and Letha Ann married, they also humbled themselves before the Lord's mighty hand. Through love, hard work, perseverance, diligence, humility, incorruptible characters, careful planning, and basic goodness to other people, they accumulated an operation that included several ranches, a

William's birthday in 1937

number of houses, considerable farm machinery, and innumerable head of sheep, goats, cows, and horses.

William and Letha Ann's life was all about work and serving others. They truly loved their work. When you love your work, all the powers and favor of both heaven and earth will assist you. The work they loved and gave themselves to blessed them again and again throughout their years. Just as Proverbs 14:23 says, "In all labor there is profit." They loved their work, and their work eventually told them its secrets. They lived life to the fullest and enjoyed no less than the abundant life in God's favor and blessing. However, on January 1, 1945, at the age of eighty-seven, William's beautiful bride, confidant, and companion of fifty-six years, Letha Ann, died and went on to receive her reward. She passed away in the San Angelo hospital after a bout with pneumonia. Out of respect and in remembrance for her many kindnesses, the Menard First Baptist Church and yard were full of her friends and family. Her obituary rightly stated, among other things, that she had "fought the good fight." This child of God had kept the faith through the trepidations of pioneer life, droughts, floods, diseases, and many other difficulties too numerous to mention. The boys said her passing took a piece out of William that he never quite got back, but he didn't quit. There was no quit in him. He drove on to accomplish much more.

## Run-In with a Cow

William lived productively for many years after the loss of his darling Letha Ann. He remained active and was still riding horses until he was eighty-one. He wouldn't have stopped then except for a problem he encountered in his dairy operation that became a tragic event.

William didn't just have a dairy, he ran the dairy and helped milk about twelve cows every day in his older years. One day a hired hand inadvertently put a cow into a stall that was not her usual one, and cows are very particular about having their own areas. After the hand started milking the cow, she became startled at the prospect of being in the wrong stall and made a sudden jump back and out of it, knocking William over as he was walking by and breaking his hip. The medical professionals in Menard didn't have all the screws, brackets, and pelvic braces they have to mend this problem today. Instead, they told him to stay in bed at the hospital for several months and hope that it might heal.

William Menzies at eighty-one

Most folk, at the time, were afraid Granddad would be like most other people who had reached his age and had a problem with a broken hip, that is, he would develop a case or two of pneumonia and be gone. Since William was very good at staying out of pictures (which was really hard to do in such a large family), the family had a problem trying to get a picture of him at that late date. They at least wanted one good picture to remember him by, so they brought him a suit coat, shirt, tie, and hat and put it on him in his hospital bed. Then they had him sit up on the side of the bed and took this picture. Notice, even with his hip broken, he looks like he is still on top of the world. You can see he still had a glimmer of love and faith in his eyes. Yet there he sat with a badly broken hip.

William's hip never fully healed, and the doctor said that he'd never ride a horse again. Well, you know that wasn't going to go unchallenged. When he was eighty-four, William decided to prove the doctor wrong. He saddled a horse, mounted up, and rode off for about forty-five minutes. When he came back and dismounted, he was in so much pain that he said he thought the doctor might have been right after all. There was too much stress on his hips with the pressure of his feet in the stirrups and the continual jolting in the seat of the saddle.

William loved his dairy and kept it until his hip was broken. Having a dairy had always meant that there was never a shortage of milk or milk products on the ranch. He also loved ice cream, and the dairy helped with that too. It seemed like the family always obliged him by having ice cream socials to celebrate his birthdays.

## William's Need for Speed and Travel

William was intrigued by airplanes and liked to fly. He flew to New York twice to see his family. He would have bought a plane if he had been just a few years younger. By the time aircraft were fully developed and in wider civilian usage, he was admittedly too old to be doing crazy eights, barrel rolls, loop-the-loops, and buzzing housetops.

### Always Thrifty

True to his Scotch heritage, William was not a spendthrift. Back when Max's family was living at the old homeplace and William was living in a retirement home in Menard, Duery (one of Max's sons) happened to ride to Menard in a car with Granddad at the wheel. He remembered how Granddad started driving down the steep hill, headed north at the south edge of town on what is called Lone Star Alley. In order to save gas, Granddad put the car in neutral, turned the engine off, and coasted as far as he

# The Latter Days

Lone Star Alley—road down into Menard

could. Most of the time he would cover ten city blocks in this glide, all the way to San Saba Street, which was the main street of town.

## William's Secrets of Longevity

As William got up in years, many wondered why he was always so healthy and was able to live so long. Just a few reasons were his healthy, hardworking, clean-living lifestyle. He never used tobacco and was almost a teetotaler. The only time he made an exception to his alcohol-free life was to have a little eggnog at Christmastime. The one bottle of whiskey he kept stashed somewhere in the house for that purpose would last for years and years of Christmases.

William also believed in watching his food intake. This was not easy to do when his wife and daughters were all such great cooks. Everything they cooked "tasted so good it would make a rabbit slap a bear," my dad said. Even when William got older and his daughters would try to give him more to eat, he would say, "I'm not working and I don't need to eat anymore." Although he enjoyed eating like everyone else, he would also say, "I don't live to eat. I eat to live." His sage advice for longevity was to "always get up from the table just a little bit hungry."

Another thing he did, and he probably didn't even know the health benefits at the time since scientists have only discovered it recently, was keep his own beehives. Until his death, he always kept a jug of honey on the kitchen table. For all but a few of those years, it was definitely local honey, because he took it from his own hives right out the back door of the homeplace. He also limited his coffee intake to one cup a day.

Finally and probably most important, he wasn't afraid of work. In fact, he loved it. The Bible says, "In the sweat of your face you shall eat bread" (Genesis 3:19). Work activates the sweat glands and sends life-giving messages to various parts of the body to send energy, build muscles, and strengthen itself to be ready for more work. Without strenuous work or activity, the body's organs and systems start shutting down. Also, without work you will not be able to enjoy your food, eat it with a clear conscience, or have enough to share with others—which is truly God's plan for man. Not only was William healthy seemingly all the time, he also had a teachable spirit.

## A Student Still at Ninety-One

When William was ninety-one he attended a meeting hosted by the officials of the Soil Conservation District in San Angelo. They had been studying the effects of low rainfall, the need for range conservation, and the effects of overstocking pastureland. There also happened to be a writer from the local newspaper there. In his article, he said that William proved "the adage (once again) that a man never gets too old to learn." According to the account in the July 28, 1946, *San Angelo Standard Times*, after the speakers had concluded their presentations, others in attendance began to ask William, the old hand, so many questions that in the end he actually wound up doing most of the instructing that day. William, a very humble man, certainly came ready enough to learn and surely did. However, those in attendance were happy to learn from the experiences—the successes as well as the failures—of the only old-timer there that day: William Menzies. He was to turn ninety-one the next day, but he wasn't about to slow down.

### Always a Contractor

When William's grandson Carl moved to San Angelo, he met an old roofer named Harrison. Upon their meeting, of course, the roofer connected with the last name and said he knew William Menzies of Menard. Harrison shared a story about some business they had conducted.

When William was in his late eighties, he negotiated with Harrison to reroof his house in Menard and insisted on a ten-year warranty on the work. Not having the proper paperwork with him at the time, and with William continuing to press for the guarantee, he finally agreed. Harrison wrote out a guarantee in longhand on a piece of paper they both signed. He never thought there would be much risk, and the guarantee would probably never come to much because William was really up in years.

Almost ten years later, though, when William was nearly a hundred years old, there was a pretty good-size hailstorm. A while after the storm, the old roofer was in the vicinity to inspect some other roofs, and he happened to walk across William's front yard. William was sitting straight up in a chair on the front porch and immediately recognized Harrison.

He said, "I thought it was about time you showed up." William was still taking care of business and got a new roof according to their agreement.

## Football Fan at Ninety-Five

William regularly attended his kids' and grandkids' football games until he was ninety-five. When he could no longer walk well enough into the stadium to comfortably sit in the bleachers, he refused to be denied. He insisted that his kids drive his car into the stadium and park it next to the sidelines. Even in his nineties, he loved his kids and he loved football.

## A Lover of Cars and Machinery

Oldsmobile first came out with the Hydramatic (no clutch) transmission in the late 1930s. William bought one of the first models, because it made it so much easier for him to drive as he got older. He was still driving a car when he was ninety-eight. Even when his eyesight weakened, he still liked to prowl around, checking out the ranching operations with his grandson Roy Jacoby. He couldn't see a lot, but that didn't stop him from enjoying the fellowship, hearing the sounds of the animals, and smelling the animals and the sweet Texas breezes.

Even in his eighties and nineties William still had his original tools as well as the old tool chest he began with in the carpenter trade. He started out in the machine end of his father's shop, and though he knew they all

My brother Steve and I (sort of) fishing the San Saba at Grassy Point

came at a cost, he never lost his interest in what machines could do and how they could improve one's efficiency. To say he was handy with tools would be an understatement. He loved his tools.

## Slow Trip to Marfa

Once when William was up in years, his daughter Letha Ann and her husband, Tom Jacoby, and their son, Ray, picked him up in Menard to go out to Marfa to see Bill and Anne Menzies. They left pretty early that day, so they stopped in Sonora for breakfast.

After they had eaten and gotten back in the car, William was riding in the backseat and Ray was driving. He was pretty much doing the speed limit, but William said, "I know this car will go faster than you are driving it." He never ceased to have a need for speed.

## San Angelo Doctor's Appointment

When William's granddaughter Willie Lee Walston was fifteen years old, she was asked to drive William from Menard to San Angelo in his Oldsmobile for a doctor's appointment. She was young and still just a wisp of a little thing, not weighing over a hundred pounds back then, but she had a license to drive and was legal to be on the road no matter if she was a small package.

*The Latter Days*

The author (the littlest one), his brother Steve, and his grandmother Ella

As short as she was, Willie Lee had a problem with the brakes while driving her dad's pickup at their ranch. Even though she sat on the edge of the front seat, her legs were too short to reach the brake pedal. One time she had to stop the truck when it was heading straight toward a gate. As usual, being unable to reach the brake pedal, she wisely pulled the wheel to the right, choosing rather to hit the gatepost instead of the gate. That proved to be the better decision of her two options.

Seeing her behind the wheel of that big Oldsmobile land yacht was surely a sight. She didn't look like much more than a dot.

The reason for the appointment was that William had a cancerous spot on the top of his head that needed treatment. Willie Lee got him safely to San Angelo in short order and without incident. When they arrived, he looked at his watch and said, "I remember when it took two days to get here." Of course he was referring to the days he had to ride a horse or a wagon or buggy in order to cover the seventy miles between Menard and San Angelo.

The doctor put some little things on the spot on William's head that looked like pencil leads, probably having to do with radiation. The physician said the spot would erupt and look terrible for a while and then it would heal. Well, he was right, because the area soon began to fester, then it healed, cleared off, and everything was all right. That's just the way they did it back then.

## Hanging Out at the Homeplace

Uncle Bill and Aunt Anne were so kind and hospitable to allow our family to stay with them just about every time we came to Menard. It was always really special just to be there. I will never forget learning to ride a horse. There is no telling how many kids learned to ride on Uncle Bill's horse Patria (which means "Fatherland" or "native country" in Spanish). I learned to ride at an early age in front of William's old homeplace. Not much later I learned to ride bareback there as well. I have ridden a lot of other horses I've liked over a lifetime, but Patria is the only horse I ever loved.

Back in the day when restaurants all over the country served sugar cubes wrapped in paper, you wouldn't believe how many I gathered up from our table. As an air force family, we traveled and ate out a lot. I saved all those sugar cubes in a shoebox to give to Patria the next time I got to see him. Every time we visited, I'd go down to the barn and rub his nose, scratch his head, look into his big brown eyes, tell him how

Billie Anne Menzies' daughters: (left to right) Pam and Cynthia Cannon atop Patria, with their dog Queenie at the homeplace

much I loved him, and feed him those sugar cubes for hours. I don't think anyone else knew about the sugar cubes, so don't tell anybody. A kid never forgets stuff like that—not even over decades. Fishing and swimming in the San Saba River down at Grassy Point were just the greatest things too.

I remember the first time Dad took my brother and me camping at Grassy Point. He showed us how to catch fish as well as how to clean and scale them. That was yucky business the first time out of the chute. He also taught us to roll the fish in cornmeal and fry them in a skillet on top of an open fire.

It was there we learned how to build a fire, and at the end of the evening, he showed us an old cowboy trick. Strangest thing, he took a shovel and covered up a perfectly good fire with dirt. That caused it to smoke somewhat, it kept the mosquitoes away all night, and it also had a benefit he hadn't yet told us about. The next morning he scraped some of the dirt off the red-hot coals, put a little more wood on it, and the fire jumped up in seconds. That dirt blanket had saved all those coals through the night. I always thought that was a pretty neat trick.

Steve and the author ride with Phillip Jacoby on Juney at the Telegraph ranch. Uncle Tom is riding Buck.

As a little boy I looked forward to visiting my great-granddad at his home in town after he retired from ranching. We were an air force family, as I said previously, and lived all over the United States and Canada, but we visited Texas and the family every chance we got. Seeing the family was always the greatest treat for us kids. No matter where I have been in the world, I always carried in my heart my love for the beautiful San Saba River, Grassy Point, and the old homeplace where all this began with William and Letha Ann. It was also where my dad was born, grew up, and lived with Granddad quite a bit. Whenever time allowed, we visited all our relatives at their ranches in Menard as well.

We were blessed as a family to live in Texas on a number of occasions when I was growing up, but never enough for my part. My brother Steve and I attended college in Texas as well. They say Texas is just a whole other country, and that's true. It is different. There is no place like it anywhere in the world. The people are so neighborly. It is the only state in the union where pickups are required to have a special license plate that says "Texas Truck."

Whenever I visited Great-Granddad (G. C.) in my younger years, he would bounce me on his knee for what seemed like hours and say, "Gid-

At William's house in Menard with Roy, Ray, Phillip, and Perry (my dad). Either Steve or the author is in the blanket.

A. J. McWilliams gives Steve (left) and the author (right) a ride on a bull.

dyup!" as if I were riding a horse. Then he would give me a big bump way up in the air, like the horse was bucking me off, and let me slide down his shin and land on his foot. There wasn't anything in the world that was more fun than that! I'd laugh and scream, and he'd keep on letting me ride.

When I was a little older, on another visit, I found a huge oak tree in the backyard to climb and a beautiful concrete pool with goldfish out on the south lawn. He always kept a big jug of honey with the honeycomb on the kitchen table too. It was always in a special crystal container and was such a beautiful sight. I confess, I got into that honey jar pretty regularly.

## Great-Granddad's Rolltop Desk

I remember the big oak rolltop desk Great-Granddad used to conduct all his business. I would sit and stare at it from time to time and look at all the slots and pigeonholes it had that were full of letters and things. Oh, I would never dare to touch it. That desk had to be the most powerful desk in the world. I was awestruck by it, thinking of how much business he must have transacted on it and all the decisions he made there to run

Great-granddad's rolltop desk

such a large ranching operation. There had to have been a whole lot of tough decisions he had to make over his lifetime at that desk. Ray Jacoby, one of his grandsons, still has that desk in his home and has put it to good use.

## Uncle Bill Makes a New Garage Door

After William got a little too old to drive, Uncle Bill Chastain from California, Great-Granddad's (G. C.) good old friend, came to visit him at his place in town. Both had gotten up in years by then. Bill had never driven an automatic transmission before and didn't know how to operate it, but he was the only one who could see well enough to do much driving that day. They got into Great-Granddad's Oldsmobile in the garage (which was a separate building behind the house), and Bill put the car in what he thought was reverse. Actually, he had it in drive. When he hit the gas pedal, the car shot forward and ran right through the garage wall. Fortunately, the bottom of the wall came loose and the top held like a hinge. This allowed the wall to swing out and up as they drove through it and the vacant lot behind it to get over to the street.

# The Latter Days

With that obstacle cleared, they drove on down to the main street, which is San Saba Street, and parked in front of the store they were going to visit. However, the excitement wasn't over yet, because they jumped the curb and stopped the car on the sidewalk. Storefront businesses lined both sides of the street at the time. Once William and Bill had finished their errands, they drove back to the house.

My grandmother, Ella Kelly (she had married a Kelly long before this), was caring for Great-Granddad back then and she had her hands full. She actually cared for both him and Letha Ann for many years before Letha Ann passed away.

## Getting Baptized

William had been faithful in church for many years. However, at the spry young age of ninety-eight, he decided that since he had never been baptized, he would. His nurse in those days was a very special, godly lady named Lovie Jones. Her husband was also a very kind man who helped Lovie minister to William. Lovie convinced him that he needed to be baptized publicly, and he said he wanted to be do it at Menard First Baptist. That he did, and the Reverend J. C. Wade baptized him.

Reverend Wade baptizing Granddad at ninety-eight

## Loving Family Relationships

William and Letha Ann had a great loving, close relationship with their children, nieces, and nephews. They loved people first of all, but they had also learned that the secret to having a full life is serving and sowing into the lives of those around them. Testimony to this is a precious letter from one of Letha Ann's nieces, Evelyn Chastain Weitenger. She grew up on the Chastain ranch immediately across the San Saba River from William and Letha Ann before the Chastain family sold it to J. H. Crawford and moved to Fort Worth.

July 4, 1954

My Dear Uncle Bill [my Great-Granddad],

Your sweet old face passes before me so often that I cannot resist the urge to tell you that you are one of the grandest characters that I have ever known. You have always stood up straight and tall and true before your brothers—man.

But now, Uncle Bill, you have done the most wonderful, the most glorious and the supreme thing. You have stood up unashamed and unafraid before your children and your children's children and sweet little ole Menard and have been counted for your Lord.

You have made the world more conscious of its Savior and, therefore, have said to all men in all walks of life that the lack of knowing Jesus as their personal Savior makes a rich man a beggar and the possession of it makes a poor man a king. It would have been a great source of joy and gladness to me to have been there when you were baptized.

As I sit here and write you, Uncle Bill, I can't refrain from running back down the highway of life to memory's lane. I see myself always galloping over across the beautiful old San Saba River to Aunt Letha Ann and your house. And if I dared to hitch my horse to the front gate post and you were up at the barn on the hill, it would not be long until you would come down. And if I were to see you coming, I would hide from you. And I shall always remember when you found me how you would take me across your checkered apron and whip me with the hair brush. But I was always alerted—and always on the lookout for a chance to get it back on you.

This one I remember. You were carrying milk out of the kitchen to the milk house. You had a five-gallon can of milk in your arms and just as you stepped out of the kitchen door I poured a quart of cold water down the back of your shirt collar.

And I can still see Aunt Letha Ann standing in the smokehouse door just killing herself laughing. I don't remember of her ever putting me up to anything but she was my audience. I would adore going to see you, Uncle Bill, on your birthday and I know it would do me a world of good too. I am so jaded and worn to a frazzle. But circumstances over which I have no control have about fenced me in and I hardly have a ghost of a chance. [Evelyn was suffering from a severe ailment at the time.]

Uncle Bill, I venture you have a wonderful nurse. To me she sounded like somebody come from heaven. Tell everybody hello for me and that I love them all. You and Aunt Letha Ann did a most wonderful job raising your children.

You two together with God streamlined, molded and equipped them with all the finest and best qualities of life—you two laid the foundation and they have finished the job—they are the salt of the earth.

Well, I shall have to say good-bye for this time my sweet, old uncle. And I surely hope I shall see you before very long.

Your loving niece,<br>Evelyn

## Family Ties Were Important

There were many cousin get-togethers during pioneer days. Family ties were everything. Before televisions, cars, movie theaters, game boxes, and iPods, family time and visiting were the most important, meaningful, and highly treasured times of life—especially in the Menzies family.

It was a great blessing for me to know Great-Granddad to some degree, praise God, because he lived so long. I was also blessed to help him do what he loved to do best: eat ice cream and cake by the San Saba River when the family celebrated his one hundreth birthday. Of course he didn't go on to receive his reward until he was 102 and I was 12. God bless him.

## A True Zeal for Life

William Menzies had a zeal and a zest for life few only dream of. He planned fearlessly, took his risks carefully, worked hard, and was certainly

Spanish goats

not afraid to operate on a large scale. His personal fortitude, self-reliance, resourcefulness, creativity, and courage were readily apparent in all that he put his hand to. In due time, the Lord added His blessing. Even so, William was always faithful in the small things, staying committed to and loving his wife, family, and God in the face of tremendous obstacles, reversals, and hardships, as well as in times of great success. He made his mark on Texas by carving out the American dream as a rancher preparing the way for civilization.

## Lessons William Taught Us

Many families came to Texas from Europe, primarily to flee religious oppression. All they wanted was to be free from tyranny, to have an honest opportunity, and to be free from government control and entanglements.

The only problem was that city folk during the Depression wanted to be dependent on the government, which ushered in countless, needless, failed government welfare and assistance programs that have multiplied over the years and led to the present-day oppression. Some have been willing to trade freedom of opportunity for false security and have reaped a harvest of tyranny as their sure reward. They gave up the best part, the

*The Latter Days*

right to struggle, which brings as its blessing the seeds of inherent greatness. These seeds can only be planted in the soils of faith, hope, risk, blood, sweat, tears, and hard work.

Texas gave immigrants something truly fantastic: the right to try, fail, learn, and try again—all without the kind of interference from government that always comes with its detrimental and deadly false support systems. The truth is, we cannot as a nation have the freedom to succeed without the freedom to fail. The kind of freedom they had in Texas during William's day is the same kind we need in America today. Freedom has always been a three-legged stool. To be completely free, it takes political, religious, and economic freedom. None can long exist without the others.

## Political Freedom

In William's day they had the privilege of entering a voting booth and drawing the curtain without fear of reprisal or intimidation from hoodlums in combat boots carrying billy clubs. What is worse is to have an attorney general who protects the hoodlums instead of the voters, as we see today. We need to go back to the time when people voted their consciences as to what was best for our country and elected who they thought could best achieve it.

Old vice on a fence post

## Religious Freedom

In Great-Granddad's day, people had the freedom to follow their conscience and to worship whatever God their faith showed them was real. They had the freedom to assemble and worship without fear of someone entering their place of worship and trying to disrupt it or opening fire on them because their particular "religion" and their "holy book" tells them that is what they are supposed to do. Faith in the true and living God is what kept these early Americans secure. It made their families strong and also prospered and guided their every step. Religion provides moorings and something little heard of today: truth. We are suffering in our society today from a dearth of truth, and that necessarily leads to an absence of trust and a breakdown of community. We see it evidenced today in a national cheating epidemic as teachers have changed their students' answers on standardized tests to ensure good scores as well as to assist in their own promotions. We see it in a president who broke twenty-three of his campaign promises in his first ninety days in office. It is nearly everywhere.

## Economic Freedom

None of what was accomplished in William's time could have been possible without economic freedom. They had the right to choose their business, their locale, whether to create their own company or not to have a company, as well as what, when, and where to invest. All this, of course, was available without an oppressive tax burden. This is all hugely important in order to have a prospering economy.

In William's day there weren't any federal taxes and there was no IRS. That didn't come about until 1913. It was a dark day when the so-called progressives gained control of the federal government. They came up with the progressive tax system and promised that the federal taxes they were about to institute would only be in place to pay off a few things. These new income taxes would be temporary and then quickly be abolished. They also said that average folks would never be asked to pay more than 1 percent and the rich no more than 7 percent. Never, never, never! They promised, right? Sorry. Within

Today's spurs

four short years the little guys were taxed up to 4 percent and the big guys to 73 percent. Welcome to the progressive, socialist tax system.

What's worse, as we all know, not only has their new taxing mechanism not been abolished like they said it would, but it has grown into an insatiable monster. From that time until today, taxes have gone up like a homesick angel, and now, in addition to federal taxes, senseless environmental restrictions abound as well. Concurrently, the economy has gone down like the *Titanic*. Socialist politicians seeking to buy votes invent one entitlement program after another to redistribute the wealth of those who work, while at the same time, they take away our religious freedoms one by one. All this, I might add, has occurred with increasing rapidity over recent years.

Never forget that the larger the government becomes, the more enslaved the people are. The smaller the government, the freer the people. Thomas Jefferson said, "Government that governs least governs best." And Ronald Reagan likewise gave us fair warning, saying, "Government isn't part of the problem; it is the problem." All these so-called socialist-progressives have succeeded in doing over the years is to build a culture of generational dependency and poverty among our people. Ben Franklin said, "The best way to help people out of poverty is to make them uncomfortable in it"—not perpetuate it.

Interestingly, the ten poorest cities in the United States with populations over 250,000, according to the US Census Bureau, 2006 American Community Survey (August 2007), all have one consistent similarity. In these cities, 24 to 32 percent of the population lives *below* the poverty level. We are talking about Buffalo, New York; Cleveland and Cincinnati, Ohio; Detroit, Michigan; El Paso, Texas; Miami, Florida; Milwaukee, Wisconsin; Newark, New Jersey; Philadelphia, Pennsylvania; and St. Louis, Missouri. These are the poorest cities in our country, and what trait do they all share? None of them has had a Republican mayor for decades. Philadelphia hasn't had one since 1952, St. Louis since 1949, Milwaukee since 1908, and Newark since 1907. El Paso has never even seen a Republican mayor.

Someone got these poor people hooked on the dole. Someone sold them into the culture of dependence, and now that is all they know and want. Abraham Lincoln warned us about this long ago. He said, "You cannot help the poor by destroying the rich. You cannot strengthen the weak by weakening the strong. You cannot bring about prosperity by discouraging thrift. You cannot lift the wage earner up by pulling the wage payer down. You cannot further the brotherhood of man by inciting class hatred. You cannot build character and courage by taking away people's initiative and independence. You cannot help people permanently by doing for them what they could and should do for themselves."

Looking for a handout has always been one of the most powerful, deftly insidious enemies of mankind. What has it done to these poor people? It took away the pride they were born with and relegated them to the station of beggars. It took their faith and drowned it with fear. It completely lost their self-respect and the spark of their initiative. It quenched the burning desire in their hearts for the individual greatness they all were born with. You see, I believe everyone was born with a genius for something, and they need free enterprise to discover it. The dole took their get up and go and made it get up and leave. It took their innate creativity and ingenuity and choked it until it died. The dole took the basic valor they were born with to look life in the face along with its hard knocks, prejudice, injustice, betrayals, up and down economies, and generally poor prospects, to laugh in life's face and say, "No matter what comes my way, my God and I can handle it. God gave me a talent, and I am going to use it to be blessed, serve others, and be a blessing." Sadly, the dole has sentenced them to sit out every day on the front porch of the government plantation (known today as government housing), looking for someone to come along and help them get their rocking chairs going.

Those who tout these entitlement programs (entitlement is just a euphemism for dependency) say that without these programs most people would be deprived of certain essential things in life. To the contrary, exactly the opposite is actually the case. Amazingly, houses can be built debt-free with knowledge, hard work, surplus materials, thrift, wisdom, and patience. I've been doing it and teaching people how to do it for years. I even have a small booklet and CD series on the subject. It is called *How to Build a Debt-Free Home*. It is really fun to own houses debt-free. The same is true of buying automobiles. With just a few hundred dollars, some know-how, and knowing where to buy used cars, just about anyone can turn their basic transportation into a money-making program. I have a small booklet on how to do this titled *How to Buy a Used Car*. Entitlement programs will never teach you how to do these things or how to take little and make much. This is what the spirit of Texas is all about.

It was these entitlement programs that caused the current financial crisis in America. Aside from that, vote-seeking politicians would have us believe that senior citizens would have to eat dog food in order to survive without government subsidies for prescription drugs. Little children, they add, would soon resemble death camp survivors. They place television ads showing granny being dumped off a cliff in her wheelchair by a heartless health care assistant as a symbol of any attempt to reform or modify Medicare. All the time, however, they know that the proposal they are attacking does not affect anyone fifty-five years or older. Reforms need to be instituted in order to save Medicare for the majority of people, whereas doing nothing would soon actually push it over the cliff. Regardless, this picture of the demise of the good life is what the mainstream media and many politicians are scaring people with so as to build their constituency. Never mind that this picture is a horrible distortion of today's debate and doesn't come close to the realities that most folk face who may temporarily be living below the so-called poverty line.

Mid-1800s spurs

Even the poorest of our neighbors, however, have cell phones, microwave ovens, flat-screen televisions, air conditioning, and cars. Perhaps some of them should reallocate a portion of their income to provide sack lunches for their children to take to school. Nobody could make a cheaper, better, or more nutritional lunch than my mother, because (unlike the government) she was frugal and she cared. Neither word can be applied to the government. The truth is that our poor neighbors are overweight on average more often than the more well-to-do.

Socialism has never worked throughout history. It makes no sense to bankrupt the entire nation in the name of helping the poor. We are spending trillions on folk who are not really poor, who could take care of themselves if they just tried. Helping the poor, however, is the happy face that politicians put on any act that expands the welfare, entitlement state so they can build their voting block. They know exactly how to play to our base nature, which always prefers an easy way out, not having to work for anything, and always trying to get something for nothing. The only problem is that, as earlier generations were told, there is no such thing as a free lunch. Somebody, somewhere has to pay for it. Undeniably, the goal of these politicians is not to help the poor but to play Santa Claus and build a dependent society of voters. Earlier generations knew that you can give a man a fish and feed him for a day, or you can teach him how to fish and feed him for a lifetime. Knowing these truths are self-evident just proves how powerful the desire is to get something for nothing and to enjoy life without work really is. Never forget, dependency cultivates votes for the worst politicians, but individual responsibility and independence makes for strong families, a growing economy, lower taxes for everyone, a healthier society, and better government.

Even worse, these poor people will never feel the triumph of personal achievement, victory over unfair odds and obstacles, or the joy of helping others along the pathway of life.

We can beg our government for privilege if we want. We can beg our government to take care of us and be our mama and our daddy, but these privileges, nanny care, Social Security, unemployment insurance, bailouts, free housing, free food, free medical care, free energy credits, and so on. come only at an enormous price from a government totally incapable and unable to deliver.

Today, Social Security is, without a doubt, the biggest Ponzi scheme ever devised in the history of mankind. Under the threat of prison time, politicians force us to pay into the Social Security trust fund, promising that our money will be there to support us when we retire. Simultaneously, these same politicians take our Social Security payments and spend it on senseless, unnecessary entitlement programs for someone else. And when we retire, they will say, "Oops. We are so sorry. The system is bankrupt, so you will have to wait until you are eighty-five to receive a monthly stipend from us."

As I put pen to paper, the post office just received a two-cent-per-stamp raise, and no sooner than the new stamps were printed, the postmaster general said that the enterprise will post a $7 billion loss this year. In addition to that, the post office is petitioning Congress to reduce their mail service from six to five days a week and plans to close or consolidate multitudes of neighborhood post offices.

Anytime the government takes over something from the private sector, not only will it never make a profit again, but there will also be at least a 30 percent increase in the cost of running it and another 20 percent increase in terms of waste and graft.

Our government has bankrupted Medicare, Medicaid, Amtrak, the post office, Social Security, the entire Treasury, and many other things too numerous to name because government is incapable of running these things. Businesses, in contrast, are accountable and judged on a daily basis for their ability to make a profit and deliver superior customer service. Government is not. It never has to be efficient or make a profit. Its operatives are judged every two to four *years,* and most of them really only know how to pass the buck, get the pork, and pass out the freebies to their constituents in order to be reelected. And of course, that only works because so many of us are foolish enough to believe their promises during election campaigns and overlook it when they never follow through on their promises. Then we believe their promises during the next election cycle and overlook their failure to deliver over and over again.

The government in and of itself owns nothing. Everything that it owns, it takes from the people, including the consent to rule. If government gives a privilege, a program, or a payment to an unemployed citizen, it must first take a liberty, a privilege, and the money from a citizen who is working and paying taxes. The primary function of government, according to the Constitution, remember, is to provide for the national defense. It is to protect us from enemies, both foreign and domestic, to protect our borders (that would be nice), bring justice, regulate interstate commerce, and guide the construction of interstate roads and communications systems as well as facilitate services like air traffic control. It is not the government's job to provide our medical treatment or to run our banks, automobile manufacturers, or insurance companies.

Socialism never has been an answer to anyone's problems. Nor is it the precursor of tyranny. It is, in fact, tyranny. Deficit spending never generated jobs or prosperity. Only the free-enterprise system can do that. Jobs and wealth are only produced by the private sector when it is allowed to produce tangible goods: things you can see, feel, and touch. The social progressives and their artificially highly paid union supporters with

Cadillac benefit packages have run off most of the real wealth-producing industries from our country.

Individualism, free enterprise, innovation, self-reliance, creativity, and dogged determination have always been the answer, and this is the very spirit that has made Texas great. These things produce. Socialism could never have won the West. It took people who had vision, who lived by principles, who were frugal, fearless, and unafraid of failure. People willing to work sixty to ninety hours per week for little or no pay.

Never forget that both socialism and free enterprise are great teachers. In just seconds, socialism, for example, can teach people how to sit down on their hands and find ten reasons why they aren't personally fit to work and ten other things they have a "right" to tell those who are still working that they need to bring them right away: free housing, free medical care, food stamps, energy credits, free education, and so forth.

Free enterprise, on the other hand, will teach self-discipline, self-reliance, and ambition, as well as generate an innate desire to improve one's talents, proficiencies, and character in order to compete. Without even knowing it is happening, free enterprise teaches folk that it is a blessing to work sixty, seventy, eighty, or ninety hours a week at something they love to do for others. Private enterprise and the free market system unleashed the creative genius and the full strength and innate skill of the American people to bring about the highest standard of living any country on this planet has ever known. Free enterprise is so powerful that there are precious few of the 195 countries currently on God's green earth where US soldiers have not spilled their blood or given their lives and treasure to secure their freedom. There are, likewise, few countries where the American industrial, inventive, creative, and economic might has not blessed their individual economies.

Colt .45 revolver

Think about it. Free enterprise yearns to be free and must be left free if it is to demonstrate its world-changing magic. When free, it is the powerful, vibrant steed that has powered the economic system that has masterfully fed and supplied our country and nearly a whole world in need. But strapping it down with the feed bag of Marxist, progressive socialism instantly drains its strength, endurance, and competitive genius and morphs it into a scrawny, bone-protruding, mangy fleabag that not only struggles to get by but can't even pull its own weight.

We all remember the events that brought down the Berlin Wall and the entire Soviet Union right behind it. Free enterprise, not government, blesses every individual citizen to be the most generous, freehanded people on earth. No country in history has ever helped the poor or given more to charity than the American people. Generosity should come from the willing heart of the individual, not the heavy, calculating hand of a political party in power with the dark, hidden motive of rewarding campaign contributors and building a constituency.

Many today wonder why Texas still prospers. Here is a little inside information.

## Texas Today

The same spirit they had in William's day is still prevalent in Texas today. It is the reason why people all over the country look to Texas as the model for state government. Three prominent economists and tax specialists, two from New York and one from California, were on Fox News Channel's *Glenn Beck Show* today as I am writing, discussing the deplorable fiscal state of our country in general. They were sharing how so many of our states, from California to Rhode Island, along with the entire federal government, are on the

A saddle made in Texas, not China

verge of bankruptcy, racking up huge annual operating deficits and mountains of debt. (The current administration is borrowing forty-one cents of every dollar it spends.) When Beck asked the panel what the states should do about it, the first suggestion was simply to model their governments by the way Texas is run. How does Texas do it?

First, Texas has no personal state income tax, so the people have more of their own money to invest and prosper. Second, the Texas state constitution requires the state legislature to set a balanced budget every two years to be in effect for the next two years. Operating deficits are not allowed. Furthermore, the state constitution is very restrictive by design so that the government has only those powers specifically granted to it. The Texas constitution has no necessary and proper clause like most states that gives them latitude to do any crazy thing a politician might dream up.

When Texas was annexed by the United States in 1845, Texas didn't ask the federal government for a bailout to assume the debt they generated during their war for independence. They chose, rather, to keep their state lands and parklands and, along with them, took responsibility for their debts. All this land has been placed under the authority of the Texas land commissioner. He manages the land and takes all the mineral, oil, and gas royalties they produce each year and for the most part puts them into the state's education and economic development funds. Thus the Economic Development Office has the money to do whatever is necessary to bring

Cows looking for a feed sack

new business and manufacturing jobs to the state. They recently lured Toyota into building a plant in San Antonio, where the city had a lot of people needing work.

Texas is also a right-to-work state. This is quite helpful because it means workers have the right to work without having to join a union or pay union dues. It lets the free market work without artificially raising pay, and thus, the cost of goods is lower. It makes Texas industries and their products better able to compete among the states as well as internationally, providing more jobs for Texans.

Texas is an employment-at-will state. This means that employers have the right to hire whoever they wish. Further, they have the right to decide, on their own, to terminate an employee at anytime with or without cause. If they don't like someone's attitude or their work, they are free to fire that employee. Poor performance can be quickly eliminated without costly, drawn-out lawsuits. Employees, likewise, have the right to quit at any time for any reason, with or without cause.

Today, half of all electricity used in America comes from coal, which causes the most carbon dioxide emissions of any fossil fuel. Texas, however, is well on its way to reducing its greenhouse gas emissions. According to the report given by Barry Smitherman, chairman of the Texas Public Utility Commission, at a Stream Energy Convention in Dallas on February 27, 2010, Texas is ready to meet the future. It now leads the nation in renewable energy, both in capacity as well as in generation. Texas is the nation's number-one producer of wind power, and this alone has significantly reduced the state's need for fossil fuels. Texas also leads in smart grid elements that allow consumers to decide when to buy their electricity each day and how to buy it at the lowest possible price. It leads in innovative clean coal technologies. It leads in natural gas shale discoveries. Texas also leads in new nuclear developments, having submitted two proposed nuclear plants to the Department of Energy, awaiting approval. Today, Texas also leads the nation in reducing carbon dioxide emissions as well as other pollutants.

In the medical field, Texas is one of only a few states that has passed tort reforms and set malpractice caps for suits against doctors. As a result, lawsuit filings in the state have dropped dramatically, malpractice insurance rates have dropped 40 percent, and instead of doctors leaving the state as they had been, recruitment is up.

Texas also just recently passed another business friendly tort reform commonly referred to as "Loser Pays." It will effectively reduce a good many frivolous, meritless, meanspirited law suits. You can still sue anyone you want to in Texas, but the party that loses a motion to dismiss their claim now has to pay the other side's attorney fees and court costs. The law also has a provision that, in some cases, will penalize the "loser" for not accepting a settlement offer that is larger than what the judge or jury actually happens to award later at the trial.

Not to be caught unawares, Texas keeps a little rainy-day money set aside as well. They have several funds totaling over eight billion dollars at the time of this writing as a cushion to meet unforeseen contingencies.

Another important point about the state government is that state senators are elected every four years. Also every four years Texans elect a lieutenant governor by statewide ballot. In addition to being second in command of the state, the lieutenant governor also serves as president of the state senate. Members of the state house of representatives are elected every two years and are led by the speaker of the house, who is elected by popular vote from among their group. Also, fiscal legislation can only originate in the house, where the members have to face the voters more frequently. Although the governor definitely has the bully pulpit, makes high-level appointments, and leads all the state agencies, including the Texas Rangers and the Texas National Guard, he still has to work with the lieutenant governor and the speaker of the house to make major changes and get things done.

They have seen to a separation of powers in Texas so things can't get bogged down politically. They don't even allow their legislature to meet under the dome but for 142 days, and even then only every two years unless there is some kind of an emergency. This is, at least to some extent, so that their politicians aren't constantly making more and more senseless laws and finding more and more ways to spend more and more taxpayer dollars. If the legislators want to stick around any longer under the dome, they just don't get paid.

The Texas judiciary is also elected by the people, so they reflect the conservative values of their constituents. This is another manifestation of the frugal living philosophy of the early Texas pioneers.

Unlike so many other states, Texas didn't ask for and wasn't looking for a federal bailout during the recent economic disaster. They rejected all the federal dollars that had strings attached and accepted only the ones that were free. After all, Texas taxpayers will have to pay for their share of the stimulus dollars the feds hand out. However, they did not want to give the feds any room to mess with Texas.

Governor Rick Perry advised Education Secretary Arne Duncan that Texas would not accept an offer of $700 million for the state's education program. In his letter to Duncan, Perry wrote, "I will not commit Texas taxpayers to unfunded federal obligations or to the adoption of unproven, cost-prohibitive national curriculum standards and tests." He also said that Texas would be foolish to accept money that had strings attached.

According to Donna Garner, who taught for decades in the state's public schools and who presently is an education policy commentator for EducationNews.org, federal health care may be in the spotlight, but there is also a huge surge for a federal takeover of the public schools. "The federal government dangled the money out in front of [the states] to come on board the Common Core Standards—that means there would be national standards all across the United States for states who join the effort," she advised. Garner also said that every state in the union—except Texas and Alaska—took this carrot. Look out, because now the feds are forming national standards writing teams. She went on to say, "The problem is these teams are not made up of classroom teachers who are working with real kids in real classrooms."

Texas also has a very effective Sunset Commission that works steadily to eliminate waste, programs, and departments that have outlived their usefulness.

The Texas system of government, insppired by its pioneers, works. It is in fact so effective that the *Wall Street Journal* reported that "45 percent of the net US job creation" from June 2009 to April 2011 was in this one state.

They take care of business in Texas, pay their bills on time, have sense enough to live within their means, and give people incentives to be self-reliant. They just know how to run a railroad.

## What William's Life Said and Still Says

William Menzies loved freedom and was looking for freedom when he left New York at the age of twenty-one. He found it on the wide, open ranges of Texas. His indomitable spirit refused to be bound up by the stone and glass caverns of big-city life. He also was a man who could deal with hardship, change, and adversity. His life uniquely spanned the Civil War, the Spanish-American War, two world wars, and the Korean War. It also, amazingly, covered the full breadth of the Industrial Revolution from mechanical threshing machines and steam-powered tractors to autos, airplanes, missiles, and Sputnik. Through it all, he left a legacy of hope for those who follow him. Any poor boy with nothing but a dream, who is not afraid of a little danger and hard work, who would put his trust in God and his shoulder to the wheel could pursue and lay hold of the impossible.

William was a great leader for many reasons, not the least of which was because all truly great leaders are made great because they fear the Lord and have a desire to humbly serve others. The wisest man who ever lived told us long ago what would happen whenever these two character qualities meet in the same person: "By humility and the fear of the Lord are riches and honor and life" (Proverbs 22:4). All these things were certainly the fruit of William's life. He was still supervising his ranching operations with the help of his sons until he was ninety-one. It was then that he turned it all over to them, giving each of his eight children more than a thousand acres of land and their fair share of his livestock. However, far more valuable than all this, he gave them his faith in God, his example, his courage, his work ethic, his character, and his good name.

Although we can learn many invaluable lessons from his life, one we certainly don't want to miss is the reflection of Proverbs 22:29: "Do you see a man who excels in his work? He will stand before kings; he will not stand before unknown men." Sometimes it is important to note what a passage doesn't say. It didn't say one had to be a rocket scientist, a neurosurgeon, or as physically strong as Charles Atlas to succeed in life. You just have to work hard and love your work.

Providence may not allow us all to be especially gifted, but we all have within our grasp the power to be diligent if we are willing to pay the price. Consistently working hard at whatever God gives our hand to do is what makes for success. Character, hard work, sweat, blood, diligence, self-reliance, and determination have made far more dreams come true than intellect, strength, or beauty. So we need to receive the work that has been placed in our hands, not as a curse, but as a gift from God. If you will learn to love your work, you will be able to enjoy it and give your full strength and mind to it. Not only that, but if you will love your work, as I already mentioned, your work will tell you its secrets. When all is said and done, it will be a great blessing to you. Neither will you want to quit doing what makes you happy when you grow older. However, God will also allow you to continue using it to help others as well.

The psalmist David wrote of godly men in Psalm 92:14: "They shall still bear fruit in old age; they shall be fresh and flourishing." This was certainly true of William Menzies.

*The Latter Days*

Mule-drawn planter

Speaking of the spirit that called men west, one of William's contemporaries, Sam Walter Foss, put to paper what that voice was saying in their day: "Bring me men to match my mountains, bring me men to match my plains, men with empires in their purpose and new eras in their brains." William was just one of the many young men in Texas who answered the spirit's call.

At the age of twenty-one, he was one of many in the Scotch, French, and German confluence who came to Texas to better their lives and subdue this land. Starting out in the sheep business, he lost five hundred out of his fledgling herd to stomach worms because of wet weather. Yet he carried on, choosing rather to go into raising much stouter animals: horses and mules. Later, going west, he settled on a ranch a few hours west of Austin on the San Saba River and married Letha Ann Chastain. Her father had come to settle in that same beautiful valley right after the unpleasantness between the North and the South. During fifty-seven years of marital bliss, William and his bride raised eight children, lived through two world wars, survived floods and droughts, and endured the screwworm plague, boll weevils, and diseases in their livestock. They overcame the Great Depression, withstood attacks by coyotes and wolves, braved the constant threat of outlaws, suffered the loss of their firstborn son, and faced many other trials of life too numerous to name. It wasn't all toil and trouble though. They had many joys and untold pleasures raising their children, building their herds, operating their dairy, their farm, and their produce service, building several ranches, and serving their community and their Lord.

It was not easy hauling lumber and materials a hundred miles one way in a wagon to build miles and miles of fences, houses, and barns, but they were willing. They had that self-reliant, never-give-up, never-say-die, dogged determination to stay with things even unto death, if that was ever required. The land was so hostile that William kept a pistol in his belt and put it under his pillow every night when he went to sleep until he died at 102. When he got up in years, someone filed the firing pin off the hammer on the pistol so he wouldn't accidentally hurt himself or anyone else. His eyesight wasn't the greatest at that time, so it didn't bother him about the firing pin. All he knew was that he needed to be ready for whatever was next. What he and Letha Ann had was the same thing all the other pioneers had: the spirit of Texas. What they had might even have come up to true grit, but they didn't know it. Their commitment to the work and the life they led burned deep in their hearts. The spirit of Texas, you see, is a deep, deep commitment of the heart.

After eighty years of ranching in Texas, William was widely known and respected as a cattleman in his community and all around the state. Upon his passing in 1957 at the age of 102, the Texas senate passed a resolution honoring him as "a pioneer ranchman" and a "leader in progressive agriculture." They said that his "life exemplifies that rapidly passing generation of rugged individualists, who have been the bulwark of our state and nation throughout our history." He lived his life as a humble servant of the Lord who always thought he could and should do more.

His attitude, example, and faith in God not only allowed his life to be a blessing to those around him, but his legacy also continues to speak encouragement into the lives of many even today.

# EPILOGUE

To say pioneering Texas was not easy in William's day is the understatement of a lifetime. In closing, let's look back at a few examples of the kind of commitment and determination we have seen in the lives of these early pioneers and consider a few new ones as well. Conquering the Texas frontier not only took a commitment of the heart but most of the time the environment required that it be a commitment unto death. Pioneer living was a life-and-death struggle that took rugged men and women willing to look the possibility of death in the face every day, and go on to do what they needed to do anyway.

From the time the early Spanish settlers came to what would later become known as Menard in the mid-1700s and the Anglos began flooding in from the East, there were horrendous battles with Indians with much loss of life on both sides. Many of the early settlers gave up and went back East, never to return, having decided that the danger was too great and the cost was far too high. Many others, however, counted the cost, stayed, and fought on. Some certainly died, but most flourished. The Indians, however, were not their only problem. At the same time these settlers were fighting the Indians from the northwest, they had to

Old iron wheel

deal with the threat of the Mexicans from the south as well as the outlaws who were all around them.

When Mexico took control of Texas from Spain, they quickly mandated that only Mexican citizens could own land. It is a fact that a good many early pioneers married Mexican women in order to own land. But the influx of Anglos from the East was a mighty, mighty torrent. It was not long before Anglos outnumbered Mexicans by about ten to one. Mexico's uneasiness about this deluge of Anglos caused them to create an increasingly oppressive government. Finally, their attempts to take the pioneers' guns led to war. They probably would have had better luck trying to take their eyeteeth than their weapons. Their guns were their only protection from the Mexicans, the outlaws, and the Indians, as well as being used to kill wild game and put food on the table for their families.

Many tragic battles ensued during the Texas war for independence. The first shots were fired at Gonzales, a battle that the Texans won handily. However, the tide immediately turned against them at the Alamo. There 150 brave Texans were surrounded by approximately 4,000 Mexicans. When given the option of abandoning the Alamo, the spirit of Texas rose up in them, and to a man, they stepped across the line William Travis drew with his sword in the sand. To a man, they committed themselves to fight at their commander's side to either victory or death. They all knew that death was the most likely of these prospects, if not a certainty. Shortly thereafter another 32 Texans somehow slipped through the enemy's lines to join the Alamo defenders. Things did not look good for those brave souls, but they were committed to the cause, to Travis, to Texas independence, and to their hope for a bright future for their posterity. In the siege that ensued, they killed over a thousand Mexicans at the cost of their own lives.

The price they paid, however, was not in vain. It is a well-proven fact that few things in life speak as loudly as the blood of the martyrs. News of the battle traveled like a prairie fire driven by high winds all across Texas, kindling the hearts of their countrymen for war. Col. James Fannin with his three hundred volunteers soon fought the Mexicans at Goliad, where they were terribly outnumbered. They were all captured and mercilessly massacred. And this tragic injustice added more fuel to the fire of Texas independence.

Undeterred and emboldened by these events, many brave Texans pressed on under the leadership of Gen. Sam Houston. Santa Anna had

# Epilogue

Dusty saddles looking for riders

just foolishly divided his army. But in the face of repeated defeats, Houston risked everything to fight the Mexicans once again, this time in a surprise attack at San Jacinto. There, in 1836, they soundly defeated the Mexican army, captured Santa Anna, and won independence for Texas.

Trouble did not, however, stay at bay for long. This time the threat came from the northeast. Precipitated primarily by the threat of the abolition of slavery, the first shots were fired by the Southern states at Fort Sumter on April 12, 1861. Calling themselves the Confederate States of America, eleven Southern states with a population (including their slaves) of half that of the North, asserted their right to secede from the Union. The spirit of Texas rested upon so many young men in this fledgling state that more of them volunteered and fought in the Civil War than from any other state in the Confederacy. Tragically, four million soldiers from both sides engaged in this epic conflict that cost the lives of more than 600,000 of the country's finest young men and women. Although bound by geography, religious beliefs, and a shared respect for free enterprise as their economic system, their lifestyles and means of producing a living were so diverse that their continued unified existence was impossible. Brothers fought brothers, fathers fought sons, and countrymen fought countrymen. Over 359,000 fell for the North and 258,000 for the South. It was our country's

Bridles in the tack shed

first total war. More died in this four-year conflict than in all the wars our country has fought put together both before and since. Though costly, it ended in the preservation of our union and settled the fact for us as a nation that all men are indeed created equal and should, above all, be free. However, we must also remember that in life as well as history, nothing is as simple as it seems. It was, however, the first war that was decided totally by the industrial might of the prevailing side.

Taming Texas took pioneers like Clara Shellenberger, who, as a teenager, was run down by a band of Indians on horseback and lanced through the shoulder with a spear. Although she survived the attack thanks to her mother's bravery, she never made a full recovery. Nonetheless, she would not give up on life, and later she married the mess sergeant at Fort McKavett. When he left her and went back to New York, she wouldn't accept defeat. She got a job in Menard as a laundress and saved what little money she made. Then she went to New York City, found her husband, and brought him back to Texas. She and her husband lived to old age and enjoyed the blessings of family life, ranching, and running a thriving business in San Antonio.

Likewise, another courageous settler who had the same spirit of Texas was James Sewell. He worked on a homestead about twenty miles

Epilogue

south of what would soon be the Menzies ranch. One day while cutting cedar posts for another pioneer in order to supply the needs of his young family, he was ambushed, shot, speared, and scalped by Indians. He was willing to pay the ultimate price to tame this fruitful land, and that price was paid in full.

Taming the frontier took men like James Bradbury Jr., my cousin's great-grandfather, who at the age of seventy-five raised a posse and went after the raiders who had just killed James Sewell only to be attacked by a much larger band. Bradbury took an arrow to the stomach and was then shot, scalped, and mutilated. James had spent his life building a ranch and trying to tame the frontier. He willingly gave his life to protect his neighbors and family.

Other brave men who possessed this same spirit of fearlessness were willing to take up the rugged life of a trail-driving cowboy. The bitter cold, the sweltering heat, and having Indian raiders on one side and cattle rustlers on the other were but a few of the hardships of driving cattle to market. Men like Andes Murchison rose to the occasion as he gladly ran many trail drives in his younger days. Later, he took his earnings and, along with his brother, Ed, started building a ranching empire in Menard County. He kept it running well until his death at ninety-seven. Neither hard work, trouble, rattlesnakes, outlaws, cattle stampedes, lightning strikes, cold weather, nor heat waves could deter him. Though he has gone, he yet lives on as a true pioneer cowboy today.

The spirit of Texas rested on many in this land, including Willie Roberts, the first male Anglo born in Menard County. Tragically, however, at the age of two Willie was stricken with polio, leaving him without the use of his legs. His handicap would have been a good excuse to sit on the sidelines of life and drown in the sorrows of self-pity, but not Willie. With no legs he still became an excellent cowhand, a broncobuster, and a courageous drover on several cattle drives. He went on with no legs to build ranches and raise large herds of horses and cattle as well as managing large numbers of beehives. Over his many years, he also raised a large family and accomplished what most men only think about—making a name for himself as a man among men.

Pioneering this land took every ranchman across Texas and fearless scientists like R. C. Bushland and E. F. Knipling, who were willing to risk their lives to fight the screwworm fly. To get the victory, they had to fight the unknown, the stench of maggots, the treatment dope, and dead animals in order to find a solution, all the while refusing to ever give up. They were willing to battle the horrible screwworm fly plague with tons of toil and much of their treasure. It took these and other scientists and nearly all of the ranchers in Texas working together, trying different solutions for seventy years to finally eradicate this deadly pest. However, they succeeded at eliminating it not only from Texas but also from the entire Southern United States and Mexico. It took a herculean effort and total dedication—but they made it happen.

Each made his unique contribution, like G. C. Menzies, grandson of William Menzies. Imbued with the same spirit of Texas, he served his fellowman as a scientist, first as a naval officer in World War II and after that by studying the habits, migration patterns, and the possibilities of rabies infection among domestic bats in deadly south Texas caves. Though he was an entomologist, he was willing to volunteer for this dangerous duty. In late 1955, however, as a result of his research, he contracted rabies and gave his life to protect and better mankind.

The spirit of Texas lived on in a boy named Chester W. Nimitz born on February 22, 1885, in Fredericksburg, Texas. His dad died before his mother gave birth, so they moved in with his grandfather, who owned the Nimitz Hotel and became quite an influence in his life. Upon graduation from high school there, Chester attended the Naval Academy. After graduating with honors in 1905, he entered the US Navy. Later in his career he attained the rank of five-star admiral and was given command of the Pacific Fleet as well as all Allied air, sea, and land forces in World War II. Chester Nimitz and his soldiers, sailors, and marines fought tooth and nail in battle after bloody battle to conquer every island in the Pacific that had been lost to the Japanese in 1941. He stayed with it until they brought home the victory for the United States. He never retreated and never took leave to see his wife even once until the war was over.

That spirit moved on to yet another great Texan. David Dwight Eisenhower was born October 14, 1890, in the town of Denison in northeast Texas, near the Red River. The family later moved to Abilene, Kansas, where he finished high school. He then put his own college dreams on the shelf and worked for two years to help his older brother Edgar finish college. Later a friend suggested he apply to the Naval Academy, which he did. He passed the entrance exam only to find out that staying out of school for two years had left him too old for admission to the academy. It looked like helping his brother had cost him his own opportunity for an

Epilogue

Old horseshoes and deer antlers in the tack shed

education. Then a senator from Kansas recommended him for appointment to US Military Academy at West Point, which he accepted. Since he was called Dwight, he decided to reverse the order of his given names when he enrolled at West Point. Once there, he failed to make the baseball team, which for him was another great disappointment. But he did make the football team as a starting linebacker and running back. Upon graduation he was commissioned a second lieutenant in the US Army and served for a number of years at camps primarily in Texas and Georgia. His assignments, however, didn't allow him to see combat or gain any command experience in World War I, which was important for advancement.

Dwight became known as Ike and had a burning desire to be a great leader. His plan to obtain that goal was to become the best follower possible. Over the ensuing peacetime years he held a number of lackluster desk jobs and then was chief military aide to Gen. Douglas MacArthur, who was then the army's chief of staff. He followed MacArthur to the Philippines, and there he also served as a military adviser to the Philippine government. He later held various staff positions in California and Texas until he was appointed chief of staff to Gen. Walter Krueger, commander of the Third Army at Fort Sam Houston in San Antonio, Texas. He later held a number of staff jobs until he was chief of staff under Gen. George C. Marshall, who noticed Ike's leadership potential. Based on Marshall's recommendation, Ike was selected by President Franklin D. Roosevelt to be the commander of the European theater of operations and later supreme commander of Allied forces of the North African theater of operations.

Smith & Wesson .38 revolver

Eisenhower received these appointments at a time when few possibly wanted either job. Germany's armies had conquered all of Europe, about half of Russia, and all of North Africa, leaving only Great Britain as a tattered, torn, battered, and beaten but undaunted Allied holdout. Germany appeared to be unstoppable. However, Ike's command soon defeated the

# Epilogue

Axis forces in North Africa and successfully invaded Sicily and the Italian mainland. Subsequently, Ike was chosen by President Roosevelt as supreme allied commander in Europe and promoted to five-star general—even in preference to Marshall.

In this new role, Eisenhower planned and led the D-day invasion of the European mainland at Normandy and continued to prosecute the war to total victory and Germany's unconditional surrender on May 8, 1945.

Because of his many military successes, the Republican Party drafted Ike in 1952 to run for the presidency. He succeeded in this, too, using the simple campaign slogan, "I like Ike." His vast experience and accomplishments, along with his other winsome attitudes, leadership practices, and extraordinary diplomatic skills, have caused him to be considered by most historians as one of the ten greatest presidents of the United States.

This spirit, which is so uniquely Texas, also came upon Audie Murphy, the son of a Texas sharecropper who was as poor as dirt. Audie was born on June 20, 1924, in the small town of Farmersville, just northeast of Dallas. As a boy he spent most of his free time hunting and learned to be good in the woods with a gun and a knife. The family was always glad to cook and augment their table with the squirrels, birds, and rabbits he brought home. Being handy with a rifle would one day soon pay off for him again.

Immediately after the Japanese attack on Pearl Harbor in 1941, Audie tried to join every branch of the service, but he was turned down on two counts: his age (he was only seventeen) as well as his height and slight build. He wouldn't give up, though, and kept right on applying. He wanted to help defend his country in the time of great national emergency. Finally, his sister signed a consent form for him to be inducted into the army. His problems weren't over though. He was flat-footed and actually passed out during close order drill in the heat of basic training at Camp Wolters, Texas. His commander tried to assign him to the cooks and bakers school, but Murphy wouldn't have it. He kept insisting on becoming a combat infantryman until his commander finally relented.

Upon graduation from basic training, Audie was shipped out immediately to Europe. There he served in combat for three years, fought in nine major European campaigns, and was wounded three times. Having joined the army as a buck private, he was promoted quickly because of his bravery, first to staff sergeant and was then given a battlefield commission as a second lieutenant. Audie received thirty-three meritorious awards for bravery, including America's highest award: the Medal of Honor. He killed over 240 enemy soldiers in various battles while simultaneously wounding and capturing a host of others. Despite the military's initial reservations, Audie Murphy became the greatest American fighting soldier of all time.

Amazingly, he was released from the service in September 1945 prior to his twenty-first birthday. Following the recommendation of a friend, actor James Cagney, Audie tried to break into the movie industry in Hollywood. His start was extremely slow and rocky. Broke and almost starving, he slept on the floor of a friend's gymnasium, but he wouldn't give up. He finally got a few bit parts. Then in 1949 he landed his first starring role. He went on to make forty-four feature films over his twenty-five year career in the movies. But despite all his success, he never forgot his Texas roots. Audie kept a ranch near Dallas to which he would return quite often. He also became quite successful as a quarter horse and thoroughbred racehorse breeder. Tragically, he died in a plane crash at the age of forty-six.

This spirit of Texas was passed on to the cotton growers who cooperated with some dedicated scientists who were working to eradicate the boll weevil, one of the most destructive insects to ever attack the United States. That frontal attack on Texas began in 1892. Although there were certainly other influences at the time, this little beetle almost single-handedly decimated the cotton industry in the South. This horde of insects cut a swath from North Carolina to Georgia and Florida and through Texas all the way to California.

Steadily and determinably, scientists in their labs worked with growers in the field to test and implement their discoveries. It took no fewer than a hundred years for them to turn the tide in the eradication of this horribly destructive beetle. Huge sums of government money as well as enormous sums contributed by the growers have allowed us to come as close as we have to controlling this insect. Cotton can now once again be grown profitably, though more work is yet to be done. Yes, it is being done and will continue to be done because these scientists and the Texas producer-run foundation that supervises the eradication program will never rest and never give up until the boll weevil is completely eradicated.

Texas has continued to grow in virtually every area over the years. It has not been without problems, however. The people of this rugged state have been attacked from every direction, both from without and within,

Epilogue 255

but the same spirit of determination and individual responsibility has always helped them overcome. It also spawned their thriving economy from the Red River to the Rio Grande and has made Texas great. That spirit has been passed along to subsequent generations that have built great cattle empires, like the vast King Ranch. This spirit flowed from the roughnecks on the Texas oilrigs to the oilmen who took great risks to build large exploration and drilling companies, huge refineries, and marketing companies. It went on to the industrialists who built burgeoning factories and to the businessmen and women who have, in the same manner, created huge distribution companies.

This spirit took root in men like H. B. Zachry, who upon graduation from Texas A&M in 1922 formed a construction company and built his first bridge in Texas. That single job started what is now a global conglomerate that includes construction, cement manufacturing, sand and gravel mining, oil and gas production, ranching, insurance, and real estate, not to mention his incredible philanthropy.

This spirit of Texas animated men like Jerry Jones, owner of the Dallas Cowboys, who was himself cocaptain of the University of Arkansas national championship football team in 1964. Jones failed at several business ventures until he began a phenomenally successful oil and gas exploration business. He purchased the Cowboys in 1989 and has helped lead them to become America's Team as well as to capture five NFL championships. With a lot of help from the state of Texas, he recently spent $1.2 billion to build the largest and finest football stadium known to man for this beloved, scrapping, pigskin-toting-and-slinging team.

This Texas spirit is not through with building new businesses either. In 2005, Rob Snyder started an electric energy marketing company in Dallas called Stream Energy. Prior to starting this company, Snyder had played soccer for Notre Dame and earned a law degree. He became very successful at leveraging buyouts on Wall Street, but at that point, wanted to do something unique. The idea he struck upon was to bring the relationship-marketing concept to the sale of energy in the recently deregulated Texas electricity market. This was quite risky; it had never been done before.

About a year prior to his start date, he filed his application with the Public Utility Commission, rented office space, hired twelve employees, and started spending millions of dollars to get everything ready. His application was looking pretty good until a jealous competitor tried to thwart it. Things generally went into a logjam, and his new employees began to wonder if he would be willing or even able to sustain these huge expenses with no income during an extended launch. They were afraid he would have to throw in the proverbial towel. His media program and everything for the launch was tied to the rollout date that had already been set, but everything required approval from the state. Then, two days prior to the deadline, the Public Utility Commission gave their approval. A year or so later, that jealous competitor was out of business, but Stream Energy was growing by leaps and bounds. In fact, Stream Energy went from zero to a billion dollars in total sales in just three years. That, my friend, is hotter than a firecracker on a West Texas prairie in the noonday sun on the Fourth of July! The company subsequently expanded into marketing natural gas in Georgia in 2008 and is still growing phenomenally. Their record in the first three years has made Stream Energy the fastest growing company in the history of business in the United States and they are still growing.

The spirit of Texas is a thing deeply rooted in its people. It has been the driving force behind building great, bustling cities like Dallas, Fort Worth, Houston, Austin, and San Antonio. However, you can also find it in the smallest businesses and on the smallest ranches, farms, and towns in the Lone Star State. It can still be found in the wide spots in the road where my family comes from, places like Lott (population 668), Menard (population 1,471), and Telegraph (population 2). In fact, Telegraph has probably never had more than eight residents, unless you count their dogs and cats.

Back in 1853, while the Honorable Jefferson Davis was serving as secretary of war, he made several creative and meaningful improvements in the military before heading up that little known but great upstart country in the land of cotton called the Confederacy. Just one of his ingenious innovations was the improvement of communications between all the frontier forts with telegraph lines. This allowed all twenty-five forts in Texas to respond immediately, en masse as a ready-reactionary force to any Indian attack or even the likely possibility at the time of an invasion by Mexico.

The area just below Menard in Kimble County, where my family has a ranch, was named Telegraph Canyon because they cut so many cedar poles there for that purpose. This little town, shown on every Texas map, has never consisted of more than one acre of land and two buildings: one is a combination general store, gas station, and post office with a residence in the back, and the other is a small commercial building. It was named Telegraph for obvious reasons, although for years, Telegraph didn't have a telegraph. And then, after a number of years, they finally got one.

For the most part, Telegraph has always been a very quiet, friendly place. However, when O. L. Freeman ran the store, he shot and killed his neighbor, Ed Fleming, for attempting to sell a small tract of land for the purpose of expanding the city. It seems that Fleming wanted to open a garage to service the horseless carriages that had started coming around there. A few weeks later, an unknown rifleman, hiding behind a barn across the road, shot and killed O. L. while he was on the front porch of the store. That little fracas pretty well quashed any further expansion of the town ever since.

A little while later when Clyde Barrow and Bonnie Parker were in the vicinity, they stopped at the Livingston ranch where they asked to buy some butter and eggs as well as sought some advice on where they might find a good campsite. Mrs. Livingston thought she knew who they were from some recent newspaper articles and gave them directions to Seven Hundred Springs up the road. Just three weeks later, the papers reported the abrupt, violent end of the infamous duo known as Bonnie and Clyde.

Telegraph in 1911

Telegraph in 2009

Other than these few incidents and an occasional holdup of the store, Telegraph is not that notorious. It's a pretty quiet, friendly place most of the time. My relatives have had a ranch there for many years. I had a great time riding horses with them and playing in the sheep pens there at a very early age. My cousins Ray and Jean Jacoby have been kind enough to host the Jacoby-Menzies family reunions there for several years.

Telegraph might not be the biggest town in Texas, but it's a good one. Rancher Coke Stevenson owned a lot of the land around Telegraph at one time. And in Texas, when it is said that a man has a lot of land, he has a lot of land. But Telegraph was good enough to be his mailing address all of his life, as well as his official residence, except for the years during World War II when the state constitution required him to live in the governor's mansion.

You see, Texas is a love story. You just can't live there for any time and not love it. There may be some who could, but I never met anyone like that. In these parts, the hats are still white and the boots are brown. They still love God, their work, their guns, their land, their animals, their freedom, and saying "Howdy!"—even to strangers.

Let us give a well-deserved tribute to William Menzies and those Texans just like him who made our lives possible as well as having made the great state of Texas what it is today. If we lose the history of where we came from, we certainly won't know where we are going. God help us to at least leave it the way we found it and, hopefully, by His grace, make it even a little better.

---

William's picture and a short account of his life as a prominent pioneer have been posted for a number of years on the walls of the Hemisfair and the Institute of Texan Cultures in San Antonio.

# TEXAS SENATE RESOLUTION NO. 68

In Memory of
Mr. William Menzies

WHEREAS, Our heavenly Father, in His infinite wisdom, on October 22, 1957 removed from his earthly labors, in the one hundred and second year of his life, William Menzies; and

WHEREAS, This pioneer Menard County ranchman was born on July 29, 1855 in Aberdeen, Scotland; and

WHEREAS, His parents came to America in 1856 and settled at Irvington on the Hudson River in New York; and

WHEREAS, Mr. Menzies learned the carpenter trade from his father, and as a young man built greenhouses at Hyde Park, the Roosevelt homestead; and

WHEREAS, Mr. Menzies came to Texas by boat at the age of twenty-one, landing at Galveston; and

WHEREAS, He moved to Port Lavaca, later to Victoria, and ranched for a time in Karnes County. In July, 1887 he moved to Menard County where he spent the remainder of his life; and

WHEREAS, In 1888, Mr. Menzies married Miss Letha Ann Chastain with whom he celebrated their golden wedding anniversary in 1938 and who preceded him in death in 1945; and

WHEREAS, Mr. Menzies was a pioneer ranchman in West Texas, being engaged in the ranch business in Menard County, for more than fifty years; and

WHEREAS, He was one of the leaders in progressive agriculture, where he pioneered the raising of sheep and goats as well as cattle; and

WHEREAS, The life of William Menzies exemplifies that rapidly passing generation of rugged individualists, who have been the bulwark of our state and nation throughout our history; and

WHEREAS, It is the desire of the Senate to pay him tribute, and to express its sympathy to the surviving members of his family, namely; four sons, Bill, Alex, Max and Walter Menzies, of Menard; three daughters, Mrs. Raymond Walston and Mrs. Irby McWilliams, of Menard; and Mrs. Tom Jacoby of Junction; nineteen grandchildren and nineteen great-grandchildren; now, therefore, be it

RESOLVED, By the Senate of Texas, that a page in the Journal be set aside in memory of William Menzies; that when the Senate adjourns today it do so in his honor; and that copies of this Resolution be sent to the members of his family. Order the official seal of the Senate.

Ben Ramsey
President of the Senate

I hereby certify that the above Resolution was adopted by the Senate on October 28, 1957, by a rising vote.

Charles Schnabel
Secretary

# BIBLIOGRAPHY

APHIS Fact Sheet: "Boll Weevil Eradication." USDA Animal and Plant Health Inspection Service. March 2007, 1–3.

Barr, Bob. "Lone Star Governor Has Credibility." *Atlanta Journal-Constitution*. June 20, 2011.

Becker, Hank. "Pioneering Agricultural Research Service Entomologist Edward F. Knipling Dies." USDA Website, News & Events, March 23, 2000, 1–2.

Birth Certificate of William Menzies. Scotland Registrar of Births. Extract June 5, 1992.

Bischofhausen, David. *Menard County*. Map and a Historical Account. 1, 1992.

Bowman, Lee. "Civil War's aftermath left West Texas settlers in peril." *San Angelo Standard-Times*. March 6, 2011.

Burnett, Frank. "Never Too Old to Learn: Menard Ranchman Student in Range Conservation at 91." *San Angelo Standard-Times*. July 28, 1946.

Center for Food Security and Public Health. *Screwworm*. Iowa State University. January 2006.

Charter of Kitchens Irrigation and Manufacturing Co. The Corporate Resolution and Articles of Agreement. April 27, 1894, 1–5.

"Chastain Ancestor Defied King of England." *Menard News*. May 5, 2011, p. 8.

Chastain, Ola. Letter to Will Chastain. March 24, 1934.

Constitution of the State of Texas. Tarleton Law Library. Jamil Center for Legal Research. University of Texas at Austin. Articles II, III & IV of the Texas Constitution, 2–4, 1876.

*Encyclopedia Britannica*. Texas, 1969. First published by a Society of Gentlemen in Scotland in 1768. S.vv., "American Civil War," 1:730; "Boll Weevil," 3:888–89; "Texas Revolution and Republic," 21:895; "Walter Reed," 19:35; "Yellow Fever," 23:882.

Ewing, William. *Nimitz: Reflections on Pearl Harbor*. 1965. Pp. 1, 2, 5.

Fisher, O. C. *It Occurred in Kimble*. Houston: Anson Jones Press, 1937.

Flores, Faye. "Bald Eagles Return to the City That Venerates Them." *Atlanta Journal Constitution*, July 5, 2009.

Fox News Network. *The Glenn Beck Show.* Topic: Progressive Taxation Starting in 1913. February 8, 2010.

Fuchs, Thomas, Dan Hanselka, Dean McCorkle, Michelle Niemeyer, and John Robinson. *The Economic Impact of Boll Weevil Eradication in Texas.* 2008. Pp. 1–19.

Hill, Simon. "Bonnie Prince Charlie and the Jacobites." Scottish History Online, 1999–2007.

Jacoby, Letha Ann. Names and Life Spans. Taken from the stones in the Sleepy Hollow Cemetery, New York, 1966.

James, Maynard G. "Truth in History: The Black-Robed Regiment." www.truthinhistory.org/the–black-robed-regiment.html.

Kanel, Michael. "Depression Rerun." *Atlanta Journal Constitution.* October 22, 2008.

Kaplan, Kim. "We Don't Cotton to Boll Weevils 'Round Here Anymore." *Agricultural Research,* February 2003, 4–6.

Knipling, E. D. "About ARS (Agricultural Research Service): The Life and Vision of Edward F. Knipling Concerning the Eradication of the Screwworm," USDA Website: last modified January 10, 2005, 1–12.

Koppel, Nathan. "Texas Legislature Approves 'Loser Pays.'" *Wall Street Journal* Law blog. May 26, 2011.

Lackey, Jerry. "Sheep Industry a Vital Part of Our Ecosystem." *San Angelo Standard-Times,* August 4, 2009.

"Legless Willy Roberts Was Top Cowhand in His Day." *West Texas Livestock Weekly.* February 10, 1949.

Ligon, Scott. *Pioneer Rest Cemetery Calendar 2011.* Information on Rev. John Ten Shilling Bell Chastain, 12.

McDonald, Archie P. *Texas: A Compact History.* Abilene: State House Press/McMurry University, 2007. Pp. 53, 57, 61,70, 80, 82, 94, 107, 133, 164, 165, 176.

Menard County Historical Society. *Menard County History: An Anthology.* 1982. pp. 24, 28, 29, 32–34, 45, 46–48, 53, 71–78, 355, 381, 415–20, 432–41, 449–51, 539–40, 674–75.

*Menard News.* "Life Story of G. W. Roberts, Jr. Shows Life in Early-Day Menard County." July 3, 2008.

———. "Texas Parks and Wildlife Names Fort McKavett State Historic Site as Park of the Month." January 4, 2007.

Menzies, Alex. Letter to Letha Ann Menzies. January 9, 1916.

———. Letter to William Menzies. August 2, 1915.

———. Personal Bible. (1870). Lists the handwritten birthdays and names of the seven children of William Menzies Sr. on the cover flap.

Menzies, Ann. *The Life of Mr. and Mrs. William Menzies.* n.d.

Menzies, Kitty Sue. *The Story of Letha Ann Chastain Menzies.* October 4, 1983.
Menzies, Max D. *Family Goes in Sheep and Goats.* n.d. pp. 1–2.
Menzies, Perry. *The Life of Perry Menzies.* November 5, 1996, 1–21.
Morney, Dr. "A Tribute to the Late George Menzies." Audio portion of Morney's television program. 1956.
National Agricultural Library. Agricultural Research Service, USDA. Screwworm Eradication Collection CD. Special Collections. April 2000.
———. USDA Screwworm Eradication Records. Screwworm Eradication Collection. Special Collections CD. Images of fly laying eggs, infested wound with larvae, larvae living in wound and adult screwworm fly, 1–2, n.d.
Obituary: Letha Ann Menzies. *Menard News.* January 1, 1945.
Obituary: William Menzies. *Menard News.* October 24, 1957.
Pierce, N. H. "Nick." *The Free State of Menard.* pp. 176–79.
Roberts, George, Sr. Letters from George Roberts to His Wife, Lucy Ann Roberts. August 18, 1886 and September 4, 1866.
Skelton, Gwen. John Ten Shilling Bell Chastain. Roots web.ancestry.com, 2006–8.
Smitherman, Barry. Speech to Stream Energy Convention in Dallas, TX, February 27, 2010. Pp. 1–26.
State of New York. Westchester County. Appraisal of the Estate of George C. Menzies, Deceased. In Relation to Taxable Transfers of Property. December 20, 1912. Pp. 1–10.
State of New York. Westchester County. Naturalization Certificate for William Menzies. October 24, 1876.
Striegler, Waldemar. "He Survived Cholera to Make Long Cattle Trails." An account of the life of Willy Roberts. Only a portion was available. N.d.
Tatum, Crystal. "Dream Come True." *Rockdale Citizen.* November 30, 2008.
Texas. *Men of Texas.* 4:3684.
Texas Boll Weevil Eradication Foundation Inc. "History of the Boll Weevil, Detection, Treatment," 2009 Program Year End Summary. www.txbollweevil.org.
"Texas Refuses Federal Education Funds." AFA Journal, www.onenewsnow.com. April 2010, 6.
Texas Senate Resolution No. 68 In Memory of Mr. William Menzies. October 28, 1957.
Thomas, David. Some Chastain Family History. deaver-chastain.org/chastain History.php. April 3, 2009.
US Census Bureau. 2006 American Community Survey. August 2007, 99, 1–3.
Waller, Matthew. "Civil War heavily shaped Fort Concho." *San Angelo Standard-Times.* March 6, 2011.
Weitenger, Evelyn Chastain. Letter to William Menzies. July 4, 1954.
Wikipedia. www.wikipedia.org. S.vv., "Bald Eagle," "Cochliomyia hominivorax (the screwworm fly)," "Boll Weevil," "Deer," "Huguenot," "Jones, Jerry"

"Mourning Dove," "Red-tailed Hawk," "Mackenzie, Ranald S.," "Rio Grande Wild Turkey," "Spanish-American War," "St. Bartholomew's Day Massacre," "Sterile Insect Technique."

William and Letha Ann Menzies 50th Wedding Anniversary. Pamphlet. December 1938.

Wyatt, Fredrica Burt. Telegraph Post Office and General Store, 1–9.

# INDEX

Adams family, 19
Alamo, 87–88, 89 (*photo*)
Alexander, Alcy (Letha Ann's mother), 40
  *See also* Chastain, Alcy
Alexander, Dave (family), 40–41
Alexander, John, 157
Andregg, Peter, 184
Anglos. *See* Texas, Anglos
anthrax, 176
Apaches, 83, 84, 85, 86–87
Archer, Mildred, 38
Armstrong, N. C., 170
Arnett family, 19
Aunt Frances (*photo*), 200
automobile, 159–61
  early (*photos*), 127, 136

bank panics, 28
barber shop (*photo*), 124
beehives, 216
Bethel (*photo*), 201
Bethell, S. F., 71
Bevans State Bank (*photo*), 123
Bevans, William, 103
Beyer, Adolph (*photo*), 126
blacksmith shop (*photo*), 126
bobcats, 144
boll weevils, 183–88
  boll weevil (*photo*), 184
  eradication strategies, 185–88
  trap (*photo*), 186
bounty hunters (*photos*), 163, 164
Bowie, Jim, 86–88, 89
Bradbury, James, 108–11
Bradbury, James, Jr., 110, 249
Bradford, George, 103
brands, 151–52
bridles (*photo*), 248
bronco busting, 12–13, 16, 132–33
  (*photos*), 12, 16, 27

Brownie (dog) (*photo*), 165
Buck (horse) (*photo*), 221
bucket (*photo*), 22
Bushland, R. C., 175, 179–81, 183, 249–50 (*photo*), 180
bushwhackers, 17, 33–34, 79, 105–7, 123
butter churn (*photo*), 48

Callan, Jim, 18, 19, 71, 122
calves (*photo*), 70
Camp San Saba, 93
Campeche, battle, 91–92
Cannon, Cynthia, (great-granddaughter of William) (*photo*), 220
Cannon, Pam, (great-granddaughter of William) (*photo*), 220
car. *See* automobile
carpetbaggers, 112
Castle Menzies, 3, 5 (*photo*)
cattle drives, 111–17
  danger of ambush, 111
cattle feuds, 118–19
Caviness, James, 109
Chamberlain, family, 19
Chastain, Alcy
  1913 dies, 54
  Letha Ann's care, 54
Chastain, Bill, 136, 138, 135 (*photo*)
  garage door, 224–25
Chastain, Clara, 21–22, 135 (*photo*), 136
Chastain, Cora, 136–37
Chastain, Evaline, 40
Chastain family
  Huguenot ancestors, 37–38
  William as boarder, 23
Chastain, Jim (brother of Letha Ann), 23, 40, 71, 157, 136
  1887, meets William, 21–22
Chastain, John, in Revolution, 38–40

Chastain, Littleton Maxwell ("L. M.") (father of Letha Ann), 37
  1858 arrives in Texas, 40
  1880s arrives in Menard County, 19
  1903 dies, 54
Chastain, Mary (sister of Letha Ann), 37 (*photo*), 40
Chastain, Maxwell (brother of Letha Ann), 40, 41
Chastain, Perry (brother of Letha Ann), 40
Chastain, Pierre, 38
Chastain, Pierre Jr., 38
Chastain ranch (*photo*), 21
Chastain, William (brother of Letha Ann), 40
Chidnster, "Buttermilk," 46–47"
Chisholm, Jesse, 112–13
Chisholm Trail, 112–17
cholera, 176–78
chuck wagon 113 (*photo*), 113–14
chute (*photo*), 59
Colt 45 revolver (*photo*), 236
Comanches, 83, 85–86, 86–87
cook, for cowboys, 113–14
cooking implements (*photo*), 47
cotton, 183–88
  cotton cart, 185 (*photo*)
  Menard County, 184
cows (*photo*), 238
coyotes, roundup, 162–64
  hunt (*photo*), 163
Crawford, Anne (*photos*), 74, 130. *See also* Menzies, Anne
Crawford, Vernon (*photo*), 198
Crockett, Davy, 89
Crowell, Sod, 175
Cutrer, Gary, *xi*

267

dairy implements (*photo*), 47
Davis, A. E., 197
DDT, 166, 187
deer, 141–42
  antlers (*photo*), 251
Delaine sheep, 64
  (*photos*), 63, 65
  ram skull (*photo*), 176
Depression, 167–68; *see also* Great Depression
Dietz, Ella (*photo, ca. 1915*), 134
Dietz, Elsie (*photo, ca. 1915*), 134
Doebler, 101
dogs, 47
Doubleday, Abner, 95
Dozier, Clyde, 170
droughts, 183

eagles, 164–67
economic freedom, 230–32
Eisenhower, David Dwight, 250–53
Eisenhower, Edgar, 250
Elder, Andy, 7–8
Ellis, Asa, 121
Ellis, Louis, 72
Ellis, Mary, 72
Esteppe, Blue, 68–70
Everett, Allan, *xi*, 165 (*photo*)
Everett, Pat, 143 (*photo*)
Everett, Willie Lee, *xi*

Faber, George, 44 (*photo*)
Fannin, James, 90
farm machinery (*photo*), 169
feed barn (*photo*), 77
feuds, 17
field hands (*photo*), 158
Findley, Dr., 53
Fisher, Glen, 163
fishing, San Saba (*photo*), 44
Five Mile Crossing, 87; *see also* Bowie, Jim
Fleming, Ed, 257
football, 204, 208, 217
Fort McKavett, 93
headquarters (*photo*), 94
Foss, Sam Walter, 243
Four-H (*photos*), 206–8
free enterprise, 236–37
Freeman, O. L., 257

Freeman, Orville, 183
freighters, 126 (*photo*), 148
Friudenberg, Herman, 59–60
  (*photo*), 60, 170

Gap Ranch, 61–62, 76
  1913 bought, 72
  1913 fenced, 119
  (*photos*), 71, 72
Garrison family, 19
goats, 75–77
  Angora, (*photo*), 76
  Spanish (*photo*), 228
God, Menzies family's faith, 136
Godfrey, Dick, 75
Godfrey, R. J., *xi*
Goliad, 90
Gossett, James, 107
Grassy Point, 22, 149 (*photo*)
  house relocated, 33
  picnics, 45 (*photo*), 46
Great Depression, 53–54; *see also* Depression
Greely, Adolphus, 95

hack (*photos*), 49
Hardee, Dick, 186
Harrison, 216–17
Harrison, Kittie Sue, 201 (*photo*)
  1933 marries Max, 201
  *See also* Menzies, Kitty Sue
Hartgraves, Henry, 17–18
hay rake (*photo*), 157
Hereford bull (*photo*), 34
hogs, 142–43
Holland family, 19
horses
  horseshoes (*photo*), 251
  in short supply, 18
  saddle pony, 116–17
  screwworm's demise, 182–83
  training horses, 148–49
  William's horses, 66–70
Houston, Sam, 90, 92
*How to Build a Debt-Free Home*, 232
*How to Buy a Used Car*, 232
Howell family, 19

ice cream freezer (*sketch*), 146

ice plant (*photo*), 145
Indian attacks, 82–83, 96–97, 108–110
  Anglos as target, 86
  battle practices, 87–88
  Indian scout (*photo*), 81
  *See also* Indians
Indians, 79, 81–88
  San Saba Valley, 81–83
  Spanish as target, 85–86
  *See also* Indian attacks
irrigation, 156–59

Jackson, Hub, 134 (*photo*)
Jackson, John, family, 19
Jacoby (*photo*), 45
Jacoby, Jean, 258
Jacoby, Lynn, 111
Jacoby, Phillip (grandson of William), 111
  (*photo, 1933*), 205
  (*photos*), 195, 221
Jacoby, Ray (grandson of William), 218, 224, 258
  (*photo, 1933*), 205
  (*photo, 1945*), 209
  (*photos*), 45, 133, 193, 195, 196, 222
Jacoby, Roy (grandson of William)
  (*photo, 1933*), 205
  (*photo, 1945*), 209
  (*photos*), 133, 193, 195, 196, 222
Jacoby, Tom, 171, 218
  (*photo, 1945*), 209
  (*photos*), 195, 221
javelinas, 142–43
Johnson, William, 121, 122
Jolly (preacher), 19
Jones, Jerry, 255
Jones, Lovie, 225
Junie (horse), 221 (*photo*)

Kansas Pacific Railway, 117 (*map*), 114–15
Kelly, Ella, 225
Kitchens Baptist Church, 57–58
Kitchens, Dick, 157
Kitchens, F. M., 19
Kitchens family, 24
Kitchens Irrigation and Manufacturing formed, 156–59
Kitchens, Lula, 23

Kitchens School, 128–30
  Class of 1908 (photo), 57
  Class of 1910 (photo), 128
  Rough Riders, 128, 129 (photo)
knife (Alex's), (photo), 139
Kniffen, John, xi
Kniffen, Katherine, xi
Kniffen, John, 81
Knipling, Edward F. "Ed," 175, 178, 179–80, 183, 186, 249–50 (photo), 180
Kothman, Charles, 170–71

Lamar, Mirabeau B., 92
Lambert, Tim, i
Landers, J. D., 18
Landers, Jakie (photo, 1948), 205
Lehmberg, W. H. (photo, 1948), 205
livestock auctions, 168–71
livestock disease
  diseases, 176–78
  screwworm, 171–83
Livingston, Jack, 102
Livingston, Mrs., 257
Lone Star Alley (photo), 215
Luckenbock Hardware store (photo), 49

Mackenzie, Randal S., 93–95
malaria, 178
Malathion, 187
McConnel, Perry, family, 40–41
McDougal, Elizabeth, 96
McDougal, William, 93, 96
McDougal's Draw, 106–7
McKavett, Henry, 93
McKinnon, Dr., 52–53
McKnight, Dr. 209
McWilliams, Alexander (A. J.) John (grandson of William)
  (photo, 1933), 205
  (photos), 197, 198, 223
McWilliams, Irby, 68, 196 (photo)
McWilliams, Menzies (M. D.) Davis (grandson of William)
  (photo, 1945), 209
  (photo, 1948), 205

(photos), 152, 197, 198, 206, 207
McWilliams, Mary Pearl (granddaughter of William)
  (photo, 1933), 205
  (photo, 1945), 209
  (photos), 45, 46, 196, 197
Meade, George Gordon, 95
Melvin, R., 175
Menard
  1911 railroad, 66
  1911 name change, 66, 123.
  See also Menard County; Menardville
Menard Auction Company, 170
Menard County, 17–18
  1858 formed, 119
  1871 organized, 119
  1972 railway station closed, 127
  churches' importance, 58
  first Anglo boy, 99–105
  first Anglo girl, 95–97
  flood of 1899, 29–34
  neighbours (photo), 135, 137
  pioneer conditions, 79
  railway celebration, 127
  railway station, 125 (photo)
  screwworm research, 178–83
  World War I, 161–62.
  See also Menard; Menardville
Menard County Road (map), 14–16
Menard Historical Society, xi
Menard, Michel B., 119
Menardville, 17
  1871 named county seat, 119
  1899 flood, 30 (photos)
  1911 name change, 66, 123
  saloons, 121–22.
  See also Menard; Menard County
Menzies, Agnes Craigmile (daughter of William), 44–45, 129, 147–48, 162
  1891 born, 192

1919 marries Walston, 192 (photos), 189, 192
Menzies, Agnes Craigmile (mother of William), 4, 6, 44–45, 46, 73
Menzies, Agnes (sister of William), d. 1869, 6
Menzies, Alex (brother of William), 6, 71–72, 138–39, 164
  1936 dies, 25–26
  lends William money, 72–73
  (photos), 26, 189
  widow's lawsuit, 167–68
Menzies, Alexander Littleton ("Alex") (son of William), 31, 53, 58, 61–62, 129–30, 175, 198–200, 204
  1899 born, 198
  1925 married, 199, 200
  (photo, c. 1915), 134
  (photo, 1945), 209
  (photos), 26, 43, 49, 52, 77, 134, 142, 190, 198, 199, 200
Menzies, Alexander Littleton ("Sonny") Jr. (grandson of William), 199, 200 (photo)
Menzies, Anne Crawford (wife of Bill), xi, 139–40, 218, 220
  1926 married, 194
  (photo, 1945), 209
  (photo), 193
Menzies, Bill, 44, 49, 55, 58, 61–62, 132, 147–48, 162, 218, 220
  (photo, 1945), 209
  (photos), 44, 147
Menzies, Billie Ann (granddaughter of William), 220
  (photo, 1929), 194
  (photo, 1933), 205
  (photos), 45, 196
Menzies, Carl (grandson of William), 175, 216–17
  (photo, 1945), 209
  (photo, 1948), 205
  (photos), 199, 200
Menzies, Charles (brother of William), d. 1874, 6

Menzies, Donna, xi
Menzies, Duery (grandson of William), 214–15
  (photo, 1945), 209
  (photos), 152, 197, 207, 208
Menzies, Duryee (brother of William), d. 1881, 6
Menzies, Ella (Pfiefer), 68, 72, 219
  first home (photo), 70
  (photo, 1945), 190, 209
Menzies family (photo, 1945), 209
  church life, 58
  coat of arms, 5
  devotion to God, 136
  faith in God, 144–45
  family tree, xv
  God's guidance, 5–6
  guns, 139–40
  motto, 5
  New York years (map), 7
  Scottish roots (map), 4
  training children, 146–49
  See also pioneer life
Menzies, Frank (brother of William), d. 1874, 6
Menzies, George (brother of William), 6, 138–39
  1912 dies, 25–26
  lends William money, 72–73
  (photo, 1890), 25
  (photos), 26, 71
Menzies, George ("G. C.") Craigmile (grandson of William), 44–45, 57, 58, 68, 72, 130–31, 190–92, 250
  1920 father dies, 191
  college, 131–32
  first home (photo), 70
  (photo, ca. 1915), 134
  (photos), 67, 73, 133, 189, 210
Menzies, George ("G. C.") Craigmile, Jr. (great grandson of William)
  1956 dies, 177
  plaque (photo), 177
Menzies, Hazel, 204
Menzies homeplace (photo), 120
Menzies ranch (photo), 31
Menzies, Iva Marie, 194 (photo)

Menzies, James, (brother of William), d. 1881, 6
Menzies, James ("Jim") William (grandson of William), 173, 175, 199, 200 (photo)
  (photo, 1945), 209
  Jim (photo), 166
Menzies, Joan (wife of Jim, photo), 166
Menzies, John Marion (grandson of William), xi, 201, 202 (photo)
Menzies, Kelly Marguerite (photo, 1945), 209
Menzies, Kitty Sue, xi
Menzies, Letha Ann (daughter of William), 111, 129, 218
  1894 born, 193
  1926 married, 195
  (photos), 138, 189
Menzies, Letha Ann ("Lee") (wife of William), xi–xii, 1867, born, 37
  1877, move to Menard County, 40–41
  1888, courtship, 23
  1888, marries William, 24
  1899, flood, 29–34
  1918–1920, first car, 159–61
  1930s, Depression, 167–68
  1938, golden anniversary, 209–10
  1945 dies, 55, 212.
  See also Menzies, Letha Ann, (photos)
Menzies, Letha Ann (photos), 39, 43, 44, 45, 46, 138, 189
  (photo, 1890), 25
  (photo, ca. 1915), 134
  (photo, 1933), 205
  (photo), 8 years old, 37
  See also Menzies, Letha Ann, business ventures
Menzies, Letha Ann, business ventures
  farm produce, 42–44, 48, 50–51
  foreclosure threat, 54
  sheep, 48, 51–52.
  See also Menzies, Letha Ann, character

Menzies, Letha Ann, character, 24, 146
  childbirth, 52–53
  devotion to church, 58
  faith in God, 211–12
  home remedies, 52–53
  love of education, 128–32
  loving relationships, 226–27
  obituary, 55
  patient with children, 133
  pioneer wife, 41–48
  wife and mother, 189
  William's greatest asset, 53–55; see also Menzies, William
Menzies, Marguerite (photo), 199
Menzies, Mary (widow of Alex, William's brother), 167–68
Menzies, Mary Louise, vii (photo, 1945), 209
Menzies, Maxwell ("Max") Duery (son of William), xi, 53, 58, 61–62, 75–76, 129–30, 132, 145, 147–48, 204
  1902 born, 201
  1933 marries, 201
  (photo, ca. 1915), 134
  (photos), 52, 165, 189, 190, 199
  sheriff, 165, 201, 202 (photo).
  See also Harrison, Kitty Sue
Menzies, Pearl (daughter of William)
  1896 born, 196
  1920 marries, 196
  (photo, ca. 1915), 134
  (photos), 138, 189
Menzies, Perry (grandson of William), vii, 57, 68–69, 136, 159–60
  (photo, 1945), 209
  (photos), 45, 133, 222
Menzies, Phillip (grandson of William) (photo), 222
Menzies, Ron, xi
Menzies, Steve (great grandson of William), 59, 222
  (photo, 1945), 209
  (photos), 218, 219, 221, 223

# Index

Menzies, Walter ("Scotty") Scott (grandson of William), 33–34, 100, 204
 *(photo, 1945)*, 209
 *(photo)*, 203
Menzies, Walter (son of William), 65, 74, 129, 162, 190
 1910 born, 203
 marries Hazel, 203 *(photo)*
 *(photo, ca. 1915)*, 134
 *(photo, 1930)*, 206
 *(photos)*, 138, 203
 tool chest *(photo)*, 9–10
Menzies, William ("Duck") Harrison (grandson of William), 201, 202 *(photo)*
Menzies, William, *xi–xii*, 6, 258–59
 1855, in Scotland 4–5
 1856, in Canada, 6
 1860, enters New York, 6
 1876, arrives Galveston, 10–11
 1878–1887, Karnes County years, 17–18
 1887, arrives in Menard, 18
 1887 buys ranch, 18
 1888, marries Letha Ann, 19, 24
 1892 Kitchens Irrigation formed, 156–59
 1899 flood, 29–34
 1913 forms Gap Ranch, 61–62
 1918–1920 first car, trucks,159
 1930s Depression, 167–68
 1938, golden anniversary, 209–10
 1945, Letha Ann dies, 212
 1957 dies (102), 244
 98, baptized, 225
 final years, 212–14, 218–20
 his life's message, 241–42
 Senate Resolution, 244, 261–62
 *See also* Menzies, William *(photos)*
Menzies, William *(photos)*, 44, 164, 189, 198, 210
 *(photo 1880s)*, 13
 *(photo, 1890)*, 25
 *(photo, ca. 1915)*, 134
 *(photo, 1933)*, 205
 *(photo, 1937)*, 211
 *(photo, 1945)*, 209
 *(photo, at 81)*, 213.
 *See also* Menzies, William, business interests
Menzies, William, business interests
 beehives, 216
 brands, 151–52
 bushwhacker threat, 33–34
 cattle, 34–35, 76
 crops, 76
 dairy, 48–50, 212–14
 Gap Ranch, 61–62
 goats, 75–77
 greenhouses (New York), 7–8
 horse racing, 29, 66–68
 horses, mules, 12–16, 16–17, 26–27, 66–70, 73–75
 industry association member, 77
 Kitchens irrigation, 156–59
 land, 71
 livestock auctions, 170
 sheep, 11–12, 17, 63–66, 76
 *See also* Menzies, William, character
Menzies, William, character, 24, 146
 devotion to church, 58
 encouraged education, 128–32, 204, 216
 faith in God 211–12
 flexible, 73–75
 football, 217
 guns, 79–81
 husband and father, 132, 189, 226–27
 interest in technology, 167
 longevity, 215–16
 love of cars, 217–18
 love of machinery, 152–59
 love of travel, 9, 214
 patient with children, 133
 progressive farmer, 156–59
 progressive rancher, 151–88
 thrifty, 214–15
 skills, 6–7
 tools, 217–18
 zeal for life, 227–228
Menzies, William Jr. (Bill) (son of William)
 1894 born, 193
 1926 married, 194
 *(photos)*, 189, 190, 193
Menzies, William Sr., *(William Menzies father)*, 4, 6
Menzies, Winston (author) (great grandson of William), *xi*, 222
 *(photo, 1945)*, 209
 *(photos)*, 60, 218, 219, 221, 223
Mergenthaler, 97
Mexico
 independence, 86–88
 Texas Anglos, 88–93
 Texas Revolution, 88–93
Miller, Bill, family, 19
Mims, Forrest, *ii*
Model T, 130 *(photo)*
 pickup, 160 *(photo)*
Moore, Edwin, 91–92
Moore, Rance, 110
Morgan (Mrs.), 19
mourning dove, 142
mule team slip *(photo)*, 156
mules, 73–75
Muller, H. J., 179–80
Munley, C. P., 19
Murchison, Allen, 122
Murchison, A. H. "Andes," 97, 99, 249–50
Murchison, Dan, 98, 99, 121
Murchison, Ed, 98, 249
Murchison, Mike, 170–71
Murchison, Mina Wilhelmina, 98
murders, unsolved, 117–18
Murphy, Audie, 253–54

Neel, Fred T., 210
Nimitz, Chester, 250
Nimitz family, 101
Noguess, Mrs. A. W., 55
Nolan, Bud, 207
Noyes, Charles, 23
Noyes, Mrs., 209

O'Bryan, Mary, 38
Open A 6 brand, 153 *(photo)*

Pacific Railway, 114–15 (*map*), 117
panthers, 143–44
Parilla, Diego Ortiz, 84
Parish, H. E., 174 (*photo*), 175
Parker, Elar, 104
pasture, 69 (*photo*)
Patria (horse) (*photo*), 220–21
Pease, Elisha M., 105
pecan trees (*photos*), 82, 83
pecans, 81–83 (*photo*), 82
Peg Leg, 105–7
Perry, Rick, 241
Pfiefer, Ella Bertha, 190–91
picnic, 124 (*photo*)
pioneer life
  church, 57–58
  get-togethers, 133–38
  ice cream socials, 145–46
  life-and-death struggles, 245–50
  murders, unsolved, 117–18
  primitive conditions, 119–21
  social life, 144–45
  threshing, 154–56
  *See also* ranch life
pioneers. *See* pioneer life; Texas, pioneers
pistol, in holster, 97 (*photo*)
political freedom, 229–30
preacher, 41; *see also* Chastain, John
Presidio de San Luis de las Amarillas, 84 (*photo*)
Puckett, Jerry, 181–82 (*photo*), 181
Puckitt, Lee Weddell, *i*

Queenie (dog), 220 (*photo*)
quilt (*photo*), 51

rabies, 177
railway
  comes to Menard, 122–28
  Pacific Railway, 114–15 (*map*), 117
  passenger travel, 138–39
  semitrailers competition, 127–28
  train in Menard (*photo*), 122
Rambo, Claude, 170
ranch life, 204–9
  chores, 132–33
  droughts, 183
  family ties, 227
  flexibility, 73–75
  industriousness, 58–61
  livestock auctions, 168–71
  machinery, 152–59
  problems, 24–25
  screwworm, 171–83
  self-reliance, 58–61
  tack, 17 (*photo*)
  tool, 64 (*photo*)
  *See also* Menzies, William, business ventures; pioneer life; sheep ranching
Reed, Walter, 178
remuda, 116
Revolution, 38–40
Reynaud, Susanne, 38
rifle, 79 (*sketch*)
robbers, 79; *see also* bushwhackers
Robbers' Roost, 80
Roberts, Florence Elizabeth, 100
Roberts, George, 100–101
Roberts, George William "Willie," Jr., 100–105, 249
  cholera, 100–102
  disability, 100, 102
  later life, 104–5
  ranch hand, 103–4
  polio, 100
Robertson, Mary, 38
Robinson, Bob, 102
Robinson, Joshua, 99–101
Robinson, Lucy Ann, 99–101
Robinson, Lucy P. D., 99–101
Robinson, Robert, 99–101
Robinson, Walter Scott, 142 (*photo*)
rolltop desk, 223–24 (*photo*)
Russel, Dick, 19
Russell, Oliver, 19
rustling, 11, 151–52

saddles
  (*photos*), 237, 247
  saddlery shop (*photo*), 125
saloons, 121–22
San Jacinto, Battle, 90–91
San Saba River, 29–34
San Saba Street, Menard, 123 (*photo*)
Santa Anna, Antonio Lopez de, 89–90, 91
Sauers, John, *i*
Saunders, 121
screwworm, 171–83
  control (*photo*), 174
  early treatments, 171–73
  fly (*photo*), 171
  fly trap (*photo*), 173
  research, 174–76, 178–83
  Stink House, 175–76
Scruggs, J. D., 24
Scruggs, Marie (*photo, 1945*), 209
Seaman, Margo, *xi*
Sears, Bob, 72
Seguin, Juan, 119
Sellers, family, 19
Sewell, James, 107–8, 248–49
Sewell, Sarah, 107–8
sheep ranching, 63–66
  decline, 163
  loss to worms, 63–64
  shearing (*photo*), 75
  sheep pen (*photo*), 10
  west Texas, 64–66
Shellenberger, Clara, 95–97, 248
Sherman, William Tecumseh, 93
shotgun
  Alex's (*photo*), 139
  double-barreled (*photo*), 141
Slaughter, Jess, 122
Smith & Wesson revolvers, 80 (*photo*)
Smith, John Alex, 18
Smitherman, Barry, 239
Snyder, Rob, 255–56
Soblet, Anne, 38
Social Security, 234
socialism, 233–36
Spanish settlements, 83–86
Spanish-American War, 1898, 28–29
spirit of Texas, 245–59
Sport (dog), 139 (*photo*), 138
Spurgeon, Dale, 187

Index 273

spurs (photo), 66
 mid-1800s, (photo), 233
stage coaches, robbery,
 105–7
 holdup (sketch), 105
Staniford, Thomas, 93
steam tractor (photo), 154
Stink House, 175–76, 183
 (photo), 175
Striegler, Arthur, 102–3
Sutton, Fred (photo, 1948),
 205
SWAHRF, 182

Tarleton State College,
 130–31
 (photo), 131
 boys (photo), 132
Taylor, 101
Teacup Mountain (photo),
 109
Telegraph
 1911 (photo), 257
 2009 (photo), 258
telegraph, 27
terracing, 156
Terreros, Fr. Alonso Giraldo
 de, 83–84
Texas
 1835 provisional
  government, 88–89
 1835–1845 republic,
  89–93
 Anglos, 86–88, 88–93,
  95–97
 carpetbaggers, 112
 Confederacy, 247–48,,
  256
 Declaration of
  Independence (of
  Texas), 119
 economy today, 237–41
 faith, 230
 freedom, 229–31
 geography, 4
 government today,
  240–41
 hunting, 138–44
 law reform, 239–40

Mexican independence,
 86–88
murders, 257
navy, 90–93
navy ensign (illus), 91
pioneers, 3–4
Spanish rule, 246–47
spirit of Texas, 245–59
State Senate Resolution,
 261–62
Texas Revolution, 88–93
trails (map), 11
U.S. annexation, 1845
U.S. government forts,
 93–97
west Texas, 63–64
Texas Boll Weevil
 Eradication Program,
 187–88
Texas longhorns (photo),
 112
Texas Rangers, 106
thresher (photo), 153
 threshing cutter (photo),
  155
 threshing machine (photo),
  155
tractor (photo), 154
trails, 17
 trail boss, 113
 trail drives, 97–98
Travis, William, 89, 246
Tumlinson, Jim, 186
turkeys. See wild turkeys
typhoid, 178

Underwood (photo), 201

Vanderstucken, Henry, 19

Wade, J. C., 225
Waits, Billy, 108
Walston (photo), 45
Walston, Agnes
 (photo, 1945), 209
 (photo), 138
Walston, Raymond Roy
 (grandson of William),
 162, 192, 193

(photo, 1933), 205
(photo, 1945), 209
(photo, 1948), 205
(photos) 45, 121, 142,
 147–48
Walston, William Raymond
 1919 marries Agnes, 192
Walston, Willie Lee
 (granddaughter of
 William), 218–19
 (photo, 1933), 205
 (photos), 32, 45, 142, 192,
 193
Ward, T. W., 105
washday equipment (photo),
 45
Watson, Marguerite, 199
 See also Menzies,
 Marguerite
Weitenger, Evelyn Chastain,
 226
welfare, 228–29
Wells, Z. M., 57
Wesrook, Ganny (photo),
 165
west Texas. See Texas, west
 Texas
Westbrook, Bud, 122
wheel, iron (photo), 245
Whitley, Hazel, 203 See also
 Menzies, Hazel
Whitley, Marion, 68
wild turkeys, 140
 turkey hunt (photo), 140
Williamson, Rusty, 47
Wilson, Fred, 18
wolves, 79, 143
 roundup,162–64
 wolf (photo), 164
woodburning stove (photo),
 42
World War I, 161–62
Worman, 145
wormies, 171
wrangler, 116

Yockey, Mrs. A. N., 105

Zachry, H. B., 255

# ABOUT THE AUTHOR

Winston Menzies is the great-grandson of William and Letha Ann Menzies and the son of Col. Perry and Mary Louise Menzies. He was born at Nellis Air Force Base, Nevada, where his father was serving as an instructor pilot during World War II. His family lived all over the United States, Canada, and many times in Texas during his dad's military career but has always maintained a ranch in Menard, Texas. He attended Tarleton State University for one year and graduated from Texas A&M University, where he also received his pilot's license and commission as a second lieutenant in the US Army.

Upon entering active duty, Winston graduated from the Airborne, Ranger, Pathfinder, and Jumpmaster schools. He served as a platoon leader in the Eighty-Second Airborne Division and was later selected to be aide to the commanding general. Subsequently he volunteered twice to command rifle companies in Vietnam. His last duty station was at Fort Ord, California, where he served as a brigade adjutant before resigning his commission as a captain to begin a business career. For his service he was awarded the Silver Star, two Bronze Stars, two Army Commendation Medals, and the Air Medal. Going to Atlanta, Georgia, for employment after his military service, his business pursuits include being founder and president of Creative Realty Inc., a commercial real estate firm; Creative Builders Inc., a building and development company; and Creative Plan Service, a plan service for builders. He later founded Creative Ministries and at the age of forty-four was called to full-time ministry and left his business interests to start living by faith. Winston has written many books and founded Creative Publishing Co. He has led many seminars and currently has an evangelistic ministry that serves a number of prisons, jails, and youth development centers in the north Georgia area. He is also the pastor of Greater Grace Church, the director of the Shepherd's House, and oversees several other outreach ministries. He and his wife, Donna Williams Menzies, have two sons, Ron and John.

Donna and Winston on their horses, Samson and Adino